PROCEEDINGS OF THE
THIRD INTERNATIONAL CONFERENCE
ON THE

internal and external

protection of pipes

LONDON, ENGLAND

SEPTEMBER 5TH-7TH, 1979

Volume 1

ORGANISED BY BHRA FLUID ENGINEERING

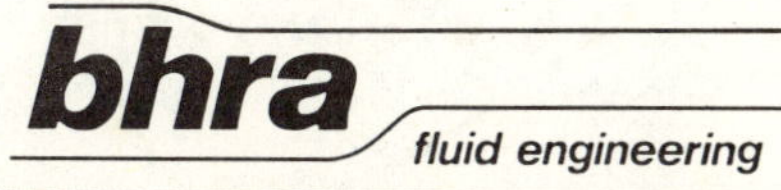

Editors: H.S. Stephens
Mrs. C.A. Stapleton

Volume 1 contains 17 of the papers presented at the
3rd International Conference on the Internal and
External Protection of Pipes. The remaining papers
and an edited record of the discussion and contributions
will appear in Volume 2.

The Organisers are not responsible for statements or
opinions made in the papers or in the discussion and
contributions.

These papers have been reproduced by offset printing
from the authors' original typescript to minimise
delay.

Printed and Published by

BHRA Fluid Engineering
Cranfield, Bedford MK43 0AJ, England

BHRA Fluid Engineering, Copyright 1979

Set of 2 Vols. ISBN 0 906085 18 7
Vol. 1 ISBN 0 906085 19 5
Vol. 2 ISBN 0 906085 20 9

ACKNOWLEDGEMENTS

The valuable assistance of the Organising Committee
and Panel of Referees is gratefully acknowledged.

ORGANISING COMMITTEE

Dr. R.V. Riley (Chairman)	Ralph V. Riley & Associates
Mr. N.G. Coles	Oyez - IBC Ltd.
Mr. G. Currer	Metal & Pipeline Endurance Ltd.
Mr. E.G. Kiernan	MK-Shand
Mr. P.R.B. Lewis	BHRA Fluid Engineering
Mr. A.T. Sneller	Watson Hawksley
Mr. H.S. Stephens	BHRA Fluid Engineering
Mrs. C.A. Stapleton	BHRA Fluid Engineering

CITATION

When citing papers from this Volume the following
reference should be used:-

Title, Author, Proc. 3rd International Conference on
the Internal and External Protection of Pipes. BHRA
Fluid Engineering, Cranfield, Bedford, England.
Volume 1, Paper No., Pages. (September, 1979).

3rd INTERNATIONAL CONFERENCE ON
THE INTERNAL AND EXTERNAL PROTECTION OF PIPES

London, England: September 5th–7th, 1979

CONTENTS

The following papers were presented:

Paper		Page
A1	The quality control of factory applied epoxy coatings to linepipe. B. Griffiths and H.S. Edward, British Gas Corporation, U.K.	1
A2	Coal tar enamels. An assessment of high temperature properties under simulated "in service" conditions. F.H. Palmer, British Petroleum Ltd., U.K., J.H. Swift, Chevron Petroleum (U.K.) Ltd., U.K., L.J. Wood, Consultant, U.K. and L. Woolf, Metrotect Ltd., U.K.	15
A3	Quality tests of plastic coated pipes for the oil and gas industry and the quality control for plastic powder for coating processes. M.H. Akstinat and W. Schuhbauer, Technical University of Clausthal, Federal Republic of Germany.	23
A4	Pipeline coating evaluation test methods and their relevance to performance. E.E. Hankins, Long Products, U.K.	47
B1	Thin film coatings for pipes and fittings into the 80's. P.A. White, Mallatite Plastics Ltd., U.K.	59
B2	Fusion bonded epoxy coatings – a review. R.F. Strobel, 3M Co., U.S.A. and B.C. Goff, 3M (U.K.) Ltd., U.K.	67
B3*	Fusion bond powder pipe coatings meeting new corrosion. challenges. W.C. Choate, Tennessee Gas Pipeline Co., U.S.A.	
B4	Recent developments in materials and methods of application of polyethylene tape pipeline coatings. G.M. Harris, The Kendall Co., U.S.A.	75
B5	Cementitious tape – A new concept in external protection of pipes and pipe insulants. R.S. Whitehouse and G.E. Richardson, Evode Ltd., U.K.	95
C1*	Advances in concrete coatings of submarine pipelines. E. Kiernan and F. E. Blake, MK-Shand, U.K.	
C2*	Splash zone protection of vertical risers using cross-linked polymeric materials. A.E.J. Strange, Crofton Coatings Ltd., U.K.	

Paper		Page

C3* Protection of downhole tubulars in North Sea production.
G. Schurch, Occidental Petroleum, U.K. and R.J. Logan, PA Inc.,
U.S.A.

C4 Polyethylene coated pipes for submarine pipeline. — 107
M. Tanaka, F. Otsuki, F. Hirano and T. Sato, Nippon Steel
Corporation, Japan.

D1 Glass piping systems and their applications. — 121
P.W. Bone, Corning Ltd., U.K.

D2 Polyurethane for the corrosion protection of pipework. — 129
J.S. Pitman, Babcock Corrosion Control Ltd., U.K.

E1 Concrete interceptor sewer corrosion protection – a state of — 139
the art report.
K.K. Kienow, Hydro Conduit Corporation, U.S.A.

E2 Use of polyamide II as pipe liner for slurry pipeline. — 157
R.A. Ganga, ATO CHIMIE, France, and V.T. Nguyen,
Omnium Technique des Transports par Pipelines, France.

E3* Organic protective coatings for concrete – and asbestos cement
pipes and structures exposed to sewage water.
T.T. Dekker, J. Happe and J. de Jong, Sigma Coatings B.V.,
The Netherlands.

G1 Safety evaluation of pipeline. — 163
D.J. Bryce and M.J. Turner, Health and Safety Executive, U.K.

G2 Accidents with pipelines in the U.S.A.: selected case histories and — 183
a review of the activities of the National Transportation Safety Board.
R.V. Riley, Ralph V. Riley & Associates, U.K.

H1 Aqueous corrosion of steel by H_2S and H_2S/CO_2 mixtures. — 205
D.E. Milliams and C.J. Kroese, Koninklijke/Shell Laboratorium,
Netherlands.

H2 A new and improved joint cover system for pipeline. — 215
K. Yamaguchi, S. Ogiwara, T. Nagasawa and I. Tsurutani,
Ube Industries Ltd., Japan.

H3* Advanced technology in polyethylene pipe systems.
H. Haartsen, and G.H.J. Van de Hoed, Stork Velsen KST,
Netherlands.

Notes:–

(i) Papers marked * will be published in Volume 2.
(ii) Session F has been allocated to the poster session.

internal and external
protection of pipes

September 5th - 7th, 1979

THE QUALITY CONTROL OF FACTORY APPLIED EPOXY COATINGS TO LINEPIPE

B.Griffiths and H.S.Edward

British Gas Corporation, U.K.

Summary

The application of sintered epoxy powder coatings to pipelines has been developing over the last decade and is now internationally recognised to be an acceptable corrosion protection with enhanced properties. In order to gain full advantage from the optimum properties of such coatings, it is necessary to achieve a very close quality control on all aspects from the basic materials through the coating processes to final inspection and handling. This requires a highly technical approach, prior discussion and approval of materials and coating procedures with comprehensive testing and overall Quality Control at all stages of production.

This paper outlines the procedures adopted by British Gas following extensive research, evaluation trials and short production runs at existing coating plants using several commercially available powders and a range of varying line pipe sizes.
These procedures include:
a) the initial evaluation and quality control tests of suitable powders
b) a coating procedure qualification to ensure efficient application/processing conditions
c) stage and final inspection of the finished product
d) supporting documentation to demonstrate that specification and Quality Control requirements
    have been achieved.

With the limited experience gained to date, it has already been demonstrated that several aspects of the process present practical difficulties, (e.g. bare pipe surface defect classification) methods of test require refinement (e.g. measurement of holidays) and further research is necessary to determine the exact parameters required of the powder and indeed which tests currently employed are relevant to powder performance in practice.

Held at Imperial College, London, England.
Organised and Sponsored by BHRA Fluid Engineering

Introduction

Although coal tar enamel and bitumen have been the traditional and relatively
reliable coatings for the corrosion protection of underground pipe systems for
many years, there exists now the committed move in the U.K. by British Gas and
other line pipe users to the application of thin film fused epoxy coating for
underground protection of pipelines.

There are several factors influencing this decision, e.g. health hazards, cost
effectiveness, repair procedures, etc., but major consideration, from the quality
control aspect, must be the comprehensive technical control that can be imposed on
the materials and the application process. Other advantages of the epoxy coating
process of pipe are as follows:

1. With the present technology, a one-coat system of sufficient thickness is
    possible

2. There exists a low loss of material where reclaim of overspray is incorporated
    into the production process

3. The speed of coating on large diameter pipes is relatively high when compared
    to other coating processes

4. Provides coatings with sufficient flexibility necessary for pipe stringing and
    bending at low ambient temperatures

5. Possesses superior adhesion to the metal substrate and provides good key for
    concrete coating

6. Bestows a good resistance to cathodic disbonding when the pipeline is cathodically
    protected.

Epoxy coatings are not susceptible to massive disbonding by general impact, however
it should be recognised that a disadvantage at present is the susceptibility to
'sharp impact' damage, therefore, care must be taken when handling and back-filling
such pipes with relatively thin coatings. In recognising this potential problem,
British Gas standards include detailed requirements for pipe handling and back-fill
materials - see BGC Standard PS/CW6 - Part 3.

Background

Liquid epoxy resins were first developed in the 1930's using amine and amide crosslinking agents but resulted in ecological problems associated with solvent release
and chemical fuming. The need to eliminate these hazards led to the introduction
of epoxy powder coatings in the early 1950's. Aided by the fluidised bed technology
of the last twenty years, powder coating is now widespread but most recent efforts to
apply thin powder coatings have been directed towards the electrostatic spraying
technique. With thin metal objects, normally protected by a comparatively thin
coating, post curing can be carried out by stoving after powder spraying, but in
the case of thick walled pipes, resin cure is by preheating of the pipes with a
short post cure if necessary.

Initially, epoxy powders were found deficient in weathering resistance and had a
high water absorption factor, however new types being developed, employing single
phase catalysed polymerisation, are overcoming these failings.

Powder Testing

In order to utilise the advantage of a high technical control on the process, it is
necessary, as a first step, to ensure that the powder material conforms closely to
the specification requirements if we are to consistently realise the optimum physical
properties in the finished coating as previously determined by laboratory evaluation.
Therefore, our first quality assurance function is with the powder manufacturer - see
BGC Standard PS/CW6 - Part 1. Once it has been determined that a certain powder
formulation is acceptable, then close quality control of all stages in its manufacture
is required in order to yield a powder of defined properties, which, on curing under
defined specific conditions, ultimately produces a coating with the desired parameters.

Laboratory formulation and testing needs to be evaluated on a full scale trial and a sufficient range of tests are necessary to fully define the requirements before the final specification for the powder can be agreed. The resin exists in three states during coating (a) powder, (b) molten mass, and (c) cured thermoset, and therefore we need specific tests relating to each of these states.

a) For powder, we have tests including particle size distribution, behaviour during electrostatic spraying, moisture content, permissible overspraying reclaim content and correctness of formulation (catalyst content).

b) In the molten state, we need to know gel time, self-levelling, edge-wetting and cure time over a range of operating temperatures.

c) With the cured coating, tests necessary are adhesion to specific steel surface profiles, flexibility, adhesion, resistance to impact damage, stress cracking, cathodic disbonding, water absorption, density and colour. It is also necessary to identify that the curing rate is being controlled to close limits and, at this time, the use of a differential scanning calorimeter (DSC) is proposed, which, by measuring glass transition temperature and change in enthalpy, can indicate conformity with desired curing rate. This DSC technique is also useful for Quality Control testing of finished coatings obtained during production to ensure that a full cure has been achieved.

With a powder quality assured by such testing, we can now examine the coating process at the applicator's plant for subsequent aspects of quality control. The coating process involves pipe surface preparation, heating, powder spraying, resin curing/cooling and final inspection including handling, storage and documentation - see BGC Standard PS/CW6 - Part 2. A schematic diagram of a typical coating plant layout is illustrated in Fig. 1.

<u>Surface Preparation</u>

a) Initial Steel Surface Quality.

The relevant British Gas line pipe specifications are BG/PS/LX4 Stage 5 for 'Seamless Line Pipe' and BG/PS/LX1 Stage 3 for 'Submerged Arc Welded Pipe'; both documents being supplementary and amending to API 5LX specification.

The main surface quality requirements of the aforementioned specifications are as follows:

(i) Laminations - extending into the face or bevel of the pipe having a transverse dimension exceeding 6.35mm are considered defects and shall be cut back until all lamination is removed.

(ii) Dents - the pipe shall contain no dents greater than 6.35mm measured at lowest point of the dent and the length of the dent shall not exceed half the pipe diameter. Cold formed dents deeper than 3.18mm with a sharp bottom shall be removed by grinding.

(iii) After blast cleaning, all slivers and scabs made visible shall be removed by grinding.

The prime purpose of the surface quality requirements of the current Line Pipe specifications is the removal of defects likely to impair the physical performance of the pipe in service, which, if allowed to remain could lead to premature mechanical or metallurgical failure.

It follows therefore that there will be a category of defects, acceptable in the aforementioned context of pipe performance that remain, including minor defects formed during the pipe processing, such as scores, dents and grit marks, and generally considered to be of a superficial nature. Our experience of previous coating techniques such as coal tar, bitumen and polyethylene cladding has justified this approach since steel quality has never posed serious problems during these external coating operations. However, epoxy coatings at the low film thickness at which they are applied, require the steel surface to have fewer imperfections and consequently inspection and grinding of defects must be more rigorous.

3

It is important to emphasise that, in the context of epoxy powder coating, surface defects illustrated on Figs. 2, 3 & 4 and minor mechanical damage, such as shown in Figure 5, are all critical and if left untreated, have been found to cause coating holidays.

These defects become more significant when the necessity to heat the pipes prior to coating is considered. Burrs or slivers, when heated to a coating temperature of around $250^{\circ}C$, are exaggerated and protrude above the pipe surface to an extent whereby holidays are created either by the steel protruding through the applied coating or by creation of a cavity in the coating on the underside of the protrusion where the powder is unable to flow before curing occurs. Figure 6 illustrates the problem.

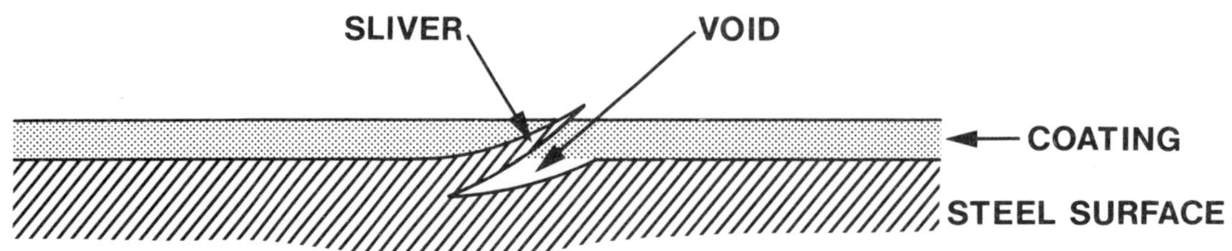

Fig. 6 Defect protruding above the pipe surface

It is necessary therefore, at the initial stage of contract negotiation, to ensure that the pipe mill is aware of the higher pipe surface standards required for epoxy coating applications.

Fortunately, these defects are easily removed on the inspection/grinding bench using an air-powered grinder with an epoxy/carborundum disc. The effect on adhesion of ground areas and epoxy smears from the grinding disc is removed by final blast cleaning.

b) Handling Contamination.

The pipe surface is liable to pick up a variety of surface contaminants during fabrication, transportation and storage prior to receipt at the applicators.

Oil and grease contamination from pipe handling equipment (during transit or in the plant), have been experienced. Because of the sporadic occurrence of this contamination and the fact that it is not readily removed by blast cleaning, visual inspection of every pipe is necessary prior to blast cleaning, and any oil or grease must be removed by washing the affected area with solvent.

c) Surface Requirements.

Experience has shown that for satisfactory adhesion, epoxy coating requires a steel surface appearance equivalent to Swedish Standard Sa $2\frac{1}{2}$ and a surface profile in the range 50 - 75 microns peak-to-trough height, with a minimum of 50 microns. Regular inspection of the blasted pipe surface condition and on shot/grit quality and cleanliness must be maintained, the blast profile being measurable by a variety of proprietary instruments.

A mixture of grit and shot has been found most satisfactory in obtaining the specified surface profile. This mixture imparts compressive stresses in the steel surface which are considered beneficial in reducing the pipe's susceptibility to stress corrosion cracking. These stresses can be measured on a laboratory-scale using an Almen gauge and work is now being carried out to develop a means of using Almen gauges under production conditions to assess these stresses.

Having attained the required surface quality by blast cleaning and grinding, it is important that good housekeeping standards are practised at the coating plant to ensure that the pipe is not subjected to further contamination. Furthermore production and quality control must ensure the pipe is coated before any rust blooming occurs.

It can be appreciated that surface quality prior to coating assumes greater
criticality for epoxy coating than for other external coating techniques, and if
sufficient attention is not paid to these details, then a variety of coating faults
ranging from holidays to total disbondment can result. With good quality control
procedures and thorough inspection, both seamless pipe and SAW pipe surfaces can be
satisfactorily prepared for coating.

d) Process Contamination.

Before any coating is commenced, it is important that any and all residual contamination is removed from the external surfaces of the pipe and this may be achieved in a
number of ways.

(i) High Pressure Water Wash.

The use of high pressure water washes have shown in the past to be very effective
in removing contamination from the surface of pipes. The water wash immediately
precedes the heating unit and consequently the risk of oxide formation is minimal.
It is important that the water is clean, neutral and not recycled.

(ii) High Pressure Air Guns.

The use of high pressure air guns is another method which is quite effective for
removing residual contaminants from pipes. Air guns are positioned close to
the surface of rotating pipe and pointing slightly upwards. They continuously
send a stream of clean dry air across the surface of the pipe.

Experience has shown that grit may also be found inside the pipe and must be
removed by high pressure air guns, otherwise the grit appears as a contaminant
in the reclaimed powder. Should the grit not be removed during the reclaim
process, excessive holidays have been found to occur on subsequent pipe
coatings.

It is also important with this cleaning method to check, for contamination,
the air used in blasting (e.g. oil and water vapour). Checks for contamination
should be carried out regularly by the use of filter paper held across the
mouth of the air gun.

(iii) Wirebrush Finishing.

On some plants, rotary wire brushes are used to remove surface contaminants.
Whilst this method can be quite effective, experience has shown that this
type of cleaning is not so successful as the previous two methods.

Heating

After thorough surface preparation, the pipe is then heated to the required
temperature for coating. The temperature required for coating will be specified
by the powder manufacturer but will generally be in the range of 230°C to 260°C and
will require to be carefully controlled within the temperature limits specified.
Several methods of heating the pipe are commercially available including:

a) Induction Coil Heating.

This is probably the most rapid method of heating, and essentially consists
of an induction coil which is energised at high frequency, producing an
electromagnetic field in the pipe which in turn generates heat. This rapid
form of heating is localised and capable of controlling pipe temperatures well
within the specified limits. To ensure that every pipe attains the required
temperature evenly throughout its length, pipe speed should be kept constant.
Pipe speed and temperature should be regularly monitored, preferably by
automatic indicators on the control panel of the induction heating unit.

b)   Indirect Heating.

This method of heating is simply an oven where pipes are stored for specific
periods of time and once optimum conditions have been established, pipe
temperatures can be achieved. The size of the oven dictates the rate of
pipes passing through it and for large diameter pipes, obviously a large
oven is required. Because of the oven size, many temperature readings at
different locations within the oven need to be monitored, if an overall
consistent temperature within the specified limits is to be obtained.

c)   Direct Heating.

In this method, the pipe passes through a refractory lined tunnel, burners
within the heater will be directed towards the pipe and refractories, thus
heating by radiation and direct flame.

This method of heating is rather more difficult to control and burner
efficiency must be carefully maintained if variation in pipe temperatures is
to be avoided. However, with modern gas control equipment, satisfactory
control can be achieved.

The methods of measuring temperatures are by temperature-indicating crayons, infra-
red detectors and bimetallic thermo-couples.

Experience has shown that whilst temperature-indicating crayons are simple to use and
can monitor temperatures effectively, care must be taken to ensure compatible epoxy
based crayons are used to avoid the possibility of creating areas of poor adhesion.

Bimetallic thermo-couples have been found not to be very satisfactory due mainly to
their slow response time and the tendency to suffer from mechanical breakdown.

Infra-red detectors are a relatively new means of temperature measurement in coating
plants and it is anticipated that this method will be used in the future.

Whichever type of temperature measurement is used, the particular equipment must be
regularly calibrated and maintained. Initially, on start up, every pipe is checked
at several places along the pipe surface. When conditions have settled down and
consistent results have been obtained, the frequency of measurement may be reduced.
The ideal situation is to have a continuous and permanent record of pipe temperatures
with a means of feedback into the heating controls for providing temperature adjust-
ments.

It is important that the pipe temperature is controlled as close to the powder spray
booth as possible, therefore, the distance between the heating source and the spray
booth should be kept to a minimum.

<u>Powder Spraying</u>

The pipe, on entering the booth is sprayed with epoxy powder using electrostatic
spray guns. The number of spray guns may vary from as little as five up to fifty
depending upon the size of spray booth, pipe diameter, spray unit, etc. The guns
may be arranged parallel to the pipe, off-set in banks or situated circumferentially
around the pipe. To date, there is no conclusive evidence to suggest that one method
is superior to another although a higher number of guns diminishes the effect of any
faulty guns and probably results in a more uniform coating thickness.

As epoxy powders are applied by electrostatic spraying, the powder must be receptive
to accepting an electrical charge. The size of the charge must be sufficient that
it will be attracted to the hot pipe surface.

One of the main physical requirements of an epoxy powder for electrostatic spraying
is the particle size and its distribution. With fine powder, guns may easily become
blocked and spraying quality may not be acceptable, therefore, the optimum particle
size distribution of the powder, relative to the particular equipment being used,
must be determined by practical experience.

6

The charge of between 60 and 100 KV is applied to the powder at the gun head, the powder being moved to the guns by air pressure of about one bar.

Resin Curing/Cooling

The powder is electrostatically attracted to a heated pipe at $240^{\circ}C + 5^{\circ}C$ (a typical coating temperature). The powder then melts and undergoes a gelling reaction. Depending upon the powder and flow agents present, the powder will flow and wet out the steel pipe surface ensuring good material contact satisfying one of our requirements (good adhesion).

The powder will remain on the pipe in this semi reacted state for about three minutes by which time the reaction will be totally completed and the pipe temperature will have dropped to about $190^{\circ}C$.

At this point, the coating is usually cold water quenched to prevent over cure and to enable the pipe to be handled without damage to the coating.

This completes the coating process and it now remains to carry out the various aspects of final inspection.

Final Inspection

With the pipe coating in a fully cured condition and all holidays detected, repaired and tested (see Appendix), it is then possible to measure the other properties expected of the coating, which are specified in BGC Standard PS/CW6 and include:

1) Thickness.

   Every pipe is checked at sufficient intervals along the pipe surface to ensure that the coating is uniform. The values obtained are critical in that too thin a coating can result in increasing handling damage, stone penetration and poor cover for metal surface defects. Conversely, too thick a coating will decrease flexibility and could give trouble during cold bending in the field.

   It is important to calibrate the thickness meter at frequent intervals and in the case of magnetic instruments, it is necessary to calibrate the instrument on the blast cleaned surface of an uncoated pipe to ensure that true readings are being taken.

2) Impact strength.

   The present method preferred by British Gas is the use of a specified weighted striker dropping through fixed distances. The resulting impact damage should be holiday free after a mean impact of 18J. Strengths of this order are obtained using a round nosed striker (B.S. 3900 - C3) but it is arguable if the results are indicative of resistance to back fill damage. Further work is considered necessary to relate a laboratory test to the normal sharp impact experienced in the field.

3) Adhesion.

   This is a difficult property to measure as the adhesion strength of epoxy coatings is too high for the normal cross-hatch method applied to paint coatings. The crude technique of scoring the coating with a strong knife merely highlights unsatisfactory adhesion, however, successful resistance by the coating to this method probably indicates an adhesion sufficient for the purpose. A further difficulty is the task of comparative study of adhesion, before and after ageing, water immersion tests, etc.

4) Hardness and degree of cure.

   Several methods have been advocated including Bucholtz indentation, solvency resistance, hot penetration probes and Differential Scanning Calorimetry (DSC)

analysis. Providing the required values for the glass-transition temperature
and enthalpy change for cured coating has been determined with precision for the
particular powder being used, then the DSC technique is accurate and rapid
enough for quality control testing.

5) Appearance.

Unless visual reference standards are retained for this test, this must be a
subjective assessment but it is considered necessary to eliminate from the
process, defects such as 'orangepeel', non-metallic inclusions, dry uncured
overspray, surface contamination, etc., all of which may give problems at a
post-inspection time. Defects may occur as a result of one or more of the
following factors:

(i)   Faulty gun operation

(ii)   Variable electrostatic charge

(iii)   Non-uniform pipe temperature

(iv)   Variable pipe speed

(v)   Inadequate extraction and recycling system

(vi)   Poor powder quality of flow or curing rate.

6) Handling and storage.

All coated pipes must be handled so as not to damage the coating. It is
considered necessary to band each pipe sufficiently with resilient material
to separate pipes from each other to prevent damage between adjacent pipes.
It is also considered worthwhile to protect the bevel ends from corrosion and
damage during transit in view of the escalating cost of bevel refurbishing.
Protection of the bevels also provides a means of minimising impact damage
(from the bevel) during handling operation.

7) Documentation.

Recognised inspection forms are maintained for all aspects of the process
including a pipe identification numbering reference in order to discriminate
throughout the records to each specific pipe processed. The forms include for
example:

(i)   Powder manufacturer's certificate of quality testing

(ii)   Steel surface defect analysis

(iii)   Record of pipe surface finish and profile

(iv)   Pipe surface temperature

(v)   Details of powder spraying variables such as number of functioning guns,
        pressures, applied voltage.

(vi)   Cure/quench time via pipe traverse speed

(vii)   Holiday count per pipe and trends

(viii) Thickness of coating per pipe

(ix)   Final inspection test results for adhesion, etc.

(x)   A record of any special pipe marking, e.g. thicker coating for concrete
        outer coating, etc.

Competent Personnel

A highly technical approach to quality control and inspection functions within the
epoxy plant is essential for achieving a satisfactory product. In meeting the
Health and Safety Executive requirements for the competence and independence of
Quality Assurance personnel, BGC policy requires supervisory staff to be qualified,
competent and well experienced chemists and/or engineers with formal training in

Supplier Quality Assurance.

For supporting inspection staff, there is a requirement for specific training and
formal assessment and certification by BGC. An approval scheme is in operation
whereby candidates undergo written, practical and oral tests to determine their
ability relative to BGC Standard PS/CW6 requirements - see BGC publication "Epoxy
Powder Coating Inspectors".

Conclusions

Experience gained by British Gas Personnel, as a result of our initial research,
evaluation trials and production runs both in the U.S.A. and Europe, leads to the
conclusion that the epoxy powder coating process provides an attractive method for
the protection of pipes used in the construction of above and below gound pipelines.
The quality control strategy for satisfactory coating operations can effectively be
achieved by ensuring that scientifically based control and test methods are applied
at all stages of manufacture from the powder production through coating applications
to final acceptance by the purchaser. To this end, it is considered that further
research and development work is needed on the application technique of epoxy powder
coatings and on the test methods to be employed in determining and controlling their
properties.

Acknowledgement

The authors wish to thank British Gas Corporation for permission to publish this
paper.

Grateful acknowledgement is also extended to the 'Non-metallics' staff, Quality
Assurance Department, H.Q., British Gas, who gave assistance in the preparation of
this paper.

References

British Gas Standard PS/CW6:

Specification for external coating for steel linepipe using resin powder systems -
    Part 1 - requirements for resin powders and methods of test for cured resin
         coatings;
    Part 2 - coating operations;
    Part 3 - field operations.

British Gas Standard PS/LX1:

Specification for submerged-arc welded linepipe 600 mm to 1400 mm nominal size
    (Supplementary and amending specification to API Spec. 5LX)

British Gas Standard PS/LX4:

Specification for seamless line pipe 150mm up to and including 450mm nominal pipe
size. (Supplementary and amending specification to API Spec. 5LX)

Swedish Standard SIS 05 59 00 - 1967

B.G.C. Publication 'Epoxy Powder Coating Approval Scheme'

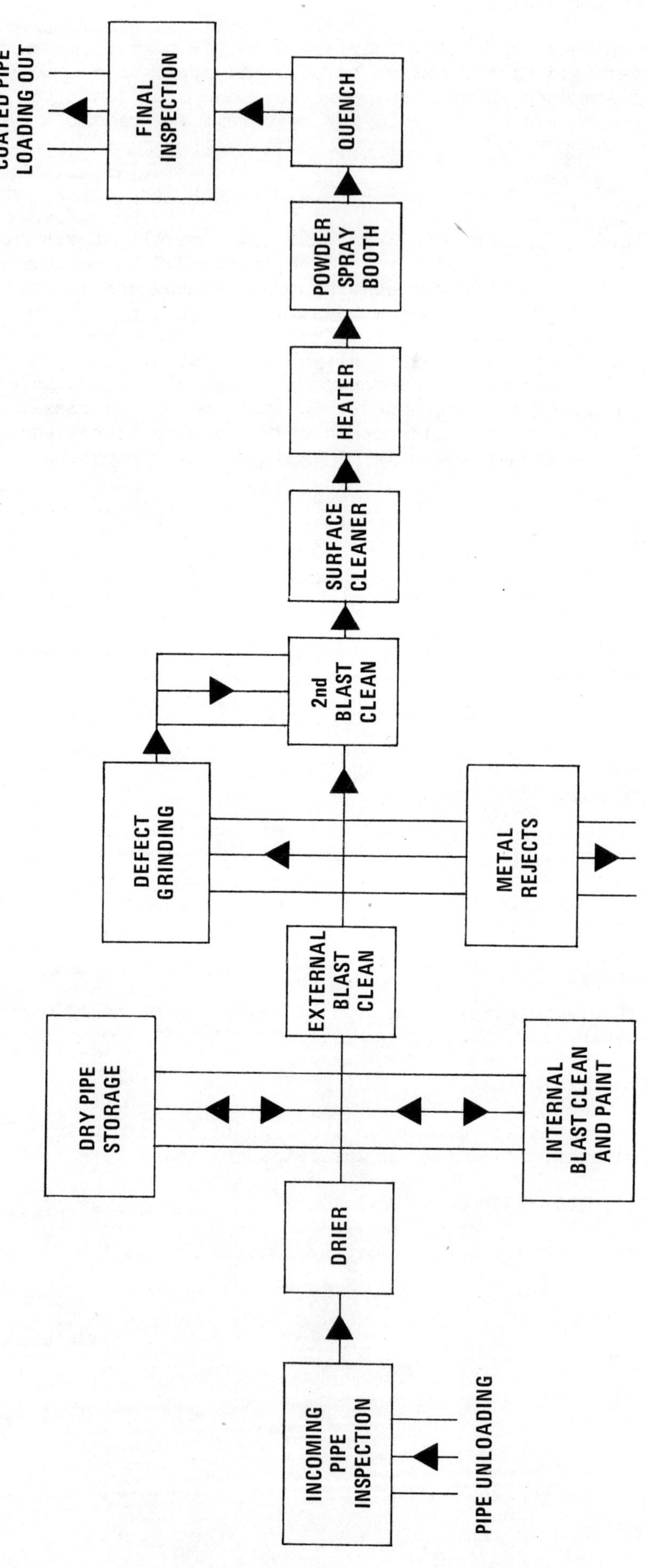

Fig. 1 Typical Epoxy Coating Plant Layout.

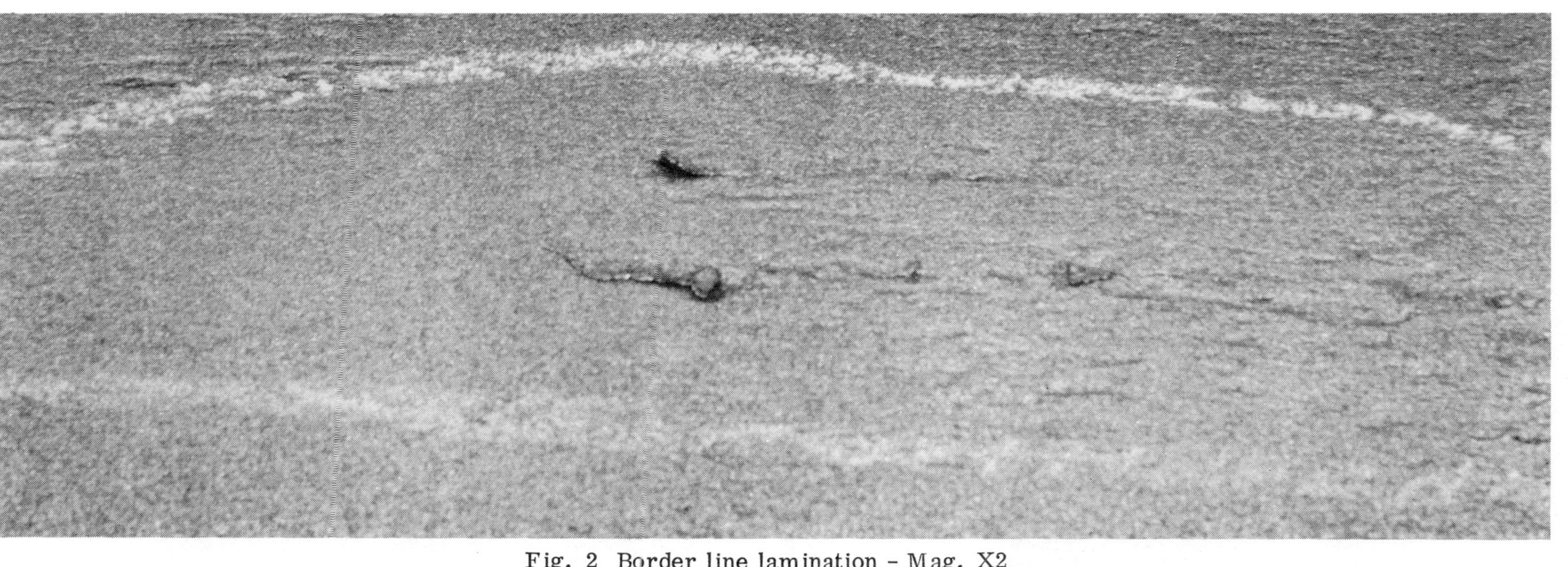

Fig. 2 Border line lamination – Mag. X2

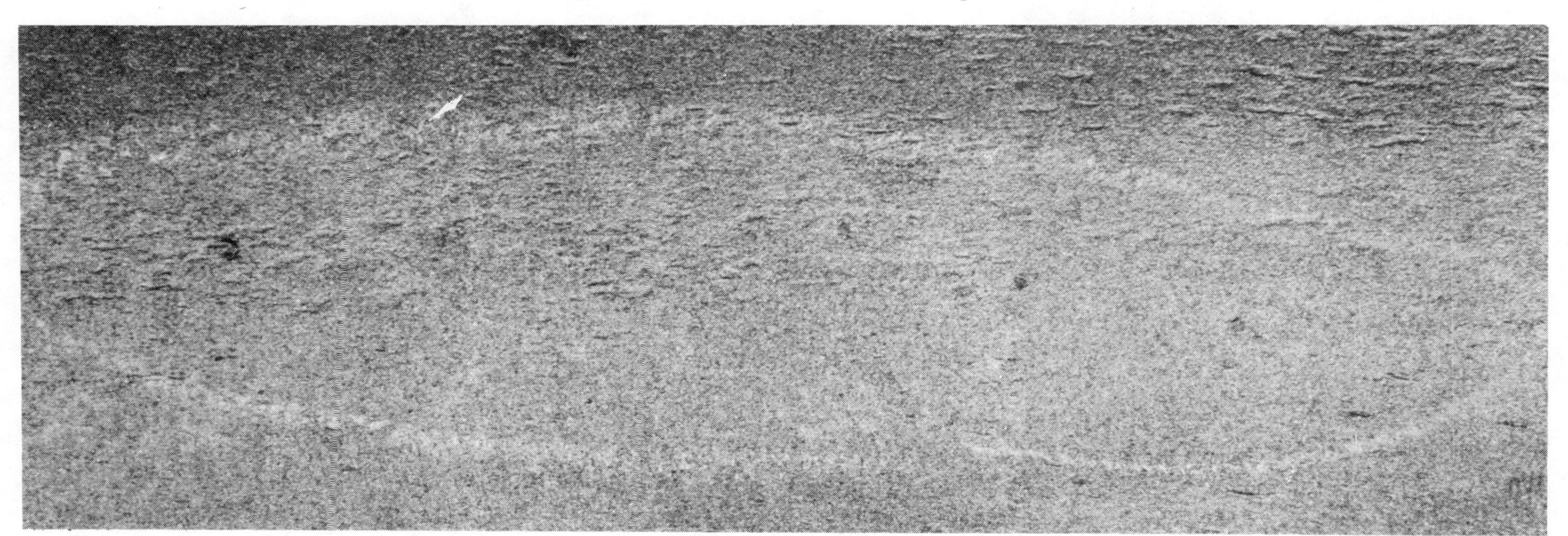

Fig. 3 Slivers and minor lamination – Mag. X1

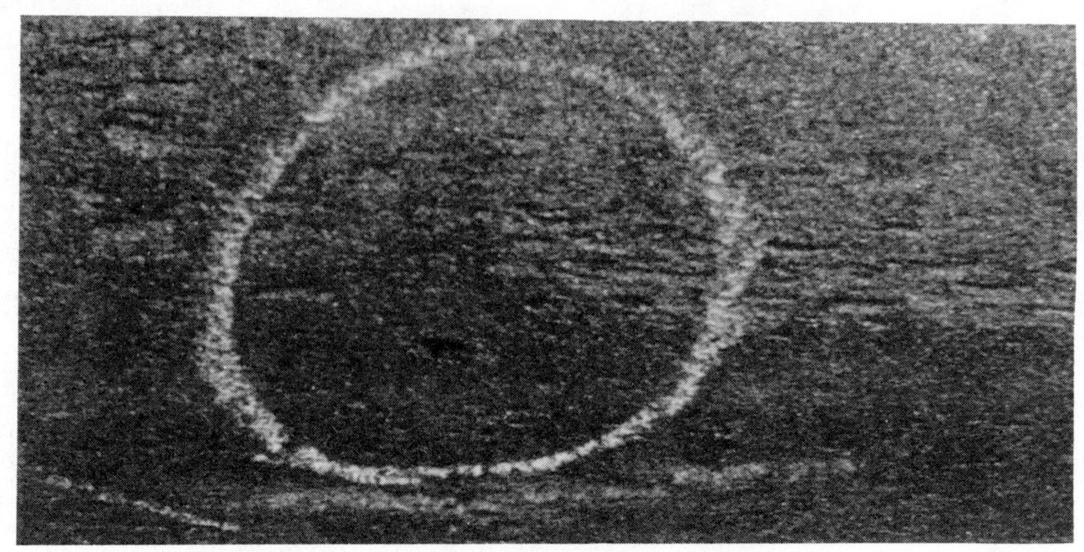

Fig. 4 Minor lamination - Mag. X1

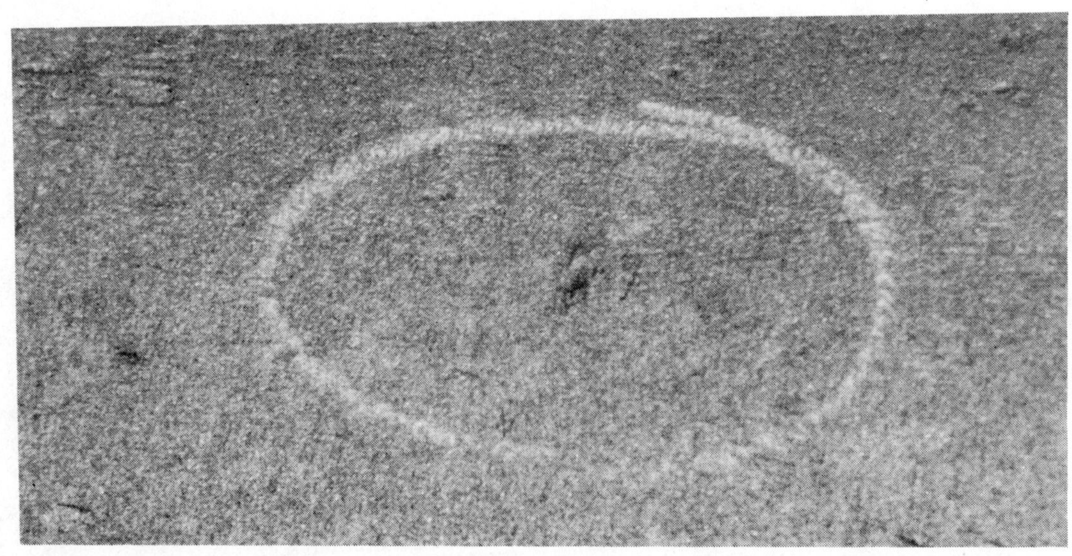

Fig. 5 Mechanical damage - Mag. X1

<u>Appendix</u>

<u>Holiday detection</u>

The ultimate criterion of successful testing is the durability of the coating under service conditions. Unfortunately, there is little information available relating holiday test results to long term performance in service and any procedures adopted will be on a largely empirical basis. It is important therefore to define procedures for holiday detection, gather all available information from results and relate to subsequent field experience.

Holidays can result when imperfections are present in the pipe coating and can occur in several ways:

(i)   Metal surface profile - with a normal distribution pattern of about a 50 micron average, 'rogue' peaks can occur.

(ii)   Particles of shot, grit or other conductive material may be imbedded into the metal surface.

(iii)   Insufficiently dressed metallic defects may remain.

(iv)   Poor coverage, flow or curing of the powder may give rise to pinholes, porosity or cavities.

The holiday test method should be determined from the detector equipment characteristics, allowable defect dimensions, coating thickness and dielectric strength of the coating.

In considering the equipment characteristics, factors affecting the selection of the probe are its geometry, conformability to pipe surface, type and durability of material. It can be shown by electric stress studies and confirmed from recent practical experience that the point-form probes (wire brush types) are more sensitive when applied to thin film coatings than on thicker films associated with coal tar, wrapping tapes, etc., and hence their use on epoxy coatings is questionable. On the aspect of conformability, the air gap between the probe and the coating is critical in avoiding corona discharge, also, if too large an air gap exists, then some holidays may remain undetected. The probe must be conductive and have a durability to enable it to survive quite severe use in service without deterioration in properties.

In deciding the voltage to be applied, the dielectric strength of the coating must be obtained. The wave form of the voltage must also be considered as too low a frequency of impulse would retard its speed of application over the coating surface but, conversely, too high a frequency produces excessive electromechanical stress levels because of the steepness of the waveform. With alternating polarity, space and charge effects may have an enhancing effect on coating breakdown, so at present, it seems likely that a continuous DC voltage will be the most controllable source.

Other characteristics required of the equipment are:

(i)   an audible and visible alarm, triggering when discharge to the pipe surface
        occurs but not when due merely to corona discharge

(ii)   sufficient screening of the leads to minimise capacitance

(iii)   appropriate safety features to protect the operator

(iv)   a practical method for calibration of the instrument e.g. a standard air gap
        electrode system.

In our experience, using a 2 KV voltage on a 400 micron coating, pulsed or continuous
DC output made no difference. However, changing from a conductive plastic blade to
a wire brush type of electrode produced a marked increase in detected holidays. This
is consistent with the theory of electrodes and may have represented unnecessary
breakdown of the coating hence the need to define all characteristics of the
equipment. The British Gas current recommended practice for holiday detection in
the coating plant is to use a conforming plastic blade with 2 KV DC source for a
400 $\pm$ 50 microns coating thickness.

internal and external
protection of pipes

September 5th - 7th, 1979

COAL TAR ENAMELS – AN ASSESSMENT OF HIGH TEMPERATURE PROPERTIES UNDER SIMULATED "IN SERVICE" CONDITIONS

F.H. Palmer, BSc.

British Petroleum Ltd., U.K.

J.H. Swift, BA (Cantab)

Chevron Petroleum (U.K.) Ltd., U.K.

L.J. Wood, PhD, CChem, FRIC, FIGasE.

Consultant, U.K.

L. Woolf, CChem, FRIC.

Metrotect Ltd., U.K.

Summary

Laboratory tests have been carried out to assess the rate and extent of creep of coal tar enamel coatings which would occur on hot oil lines in the North Sea.
The importance of selecting the correct grades of primer and enamel has been established.
The effects of glass reinforcement and also of operating temperature have been studied.

Held at Imperial College, London, England.
Organised and Sponsored by BHRA Fluid Engineering

1. GENERAL

In this programme of creep testing the experimental rig was designed as far as
possible to be a "scale down" version of operational conditions.

After shotblasting and priming 4 foot lengths of $2\frac{1}{4}$" (60mm) O.D. pipe were coated
with synthetic primer and "Hi-temp" coal tar enamel, reinforced with inner and
outer wraps of glass fibre tissue.

Hot oil was then circulated from an external thermostatically controlled tank via a
high capacity pump. Temperature measurements on test pipes were made by
mounting thermocouples on the pipe wall and also on the outer surface of the
corrosion coating.

Creep measurements were made by means of dial gauges fixed between the pipe and
concrete weight coating. Four series of creep tests have already been carried out
varying load characteristics and temperature, and also the grade of coal tar enamel.
In one series the reinforcement was omitted from the coating.

Further tests are in progress in order to establish the role of reinforcement materials
within the coating and also to establish the maximum "in service" temperature for
coal tar enamels under offshore conditions.

2. PIPELINE CONSTRUCTION

It is common practice for offshore steel pipelines conveying either oil or gas to be
protected from corrosion by an external coating of coal tar enamel several
millimetres in thickness. A cathodic protection system is normally installed. In
the case of the Ninian Field pipelines this is provided by annular sacrificial anodes
at intervals along the pipeline. Adequate negative buoyancy is achieved by casting,
or spraying, a weight coat of concrete about 40 to 60mm in thickness over the coal
tar enamel so that the external diameter of the concrete is the same as the external
diameter of the anodes. This is desirable so that the composite pipe is of constant
external diameter and will pass evenly over the stinger of the lay-barge and also so
that the anodes are not masked by concrete. Frequently crack inducers in the form
of annular sawn or cast slots are provided in the concrete at intervals of about one
metre and filled with a compressible material. At the field joints in the steel pipe
the concrete is omitted over a length of some 300mm and the gap filled with a poured
bitumen or coal tar product.

Once on the sea bed the pipeline is "buried" by either fluidising or jetting the soil
under the pipe to form a trench. The crown of the pipe is then left either level with
or below the adjacent sea bed level. This trench is then assumed to be back-filled
over a period of time by a process of natural accretion. There is no positive
compacted back-filling as would occur on land.

3. CREEP

Oil temperatures encountered in the North Sea are higher than those generally found
hitherto. These higher temperatures may approach the ring and ball softening point
of the coal tar enamel corrosion coating. For long pipelines the expansion of the
steel pipe caused by this hot oil is resisted by the overall resistance of the pipeline to longitudinal movement and is therefore taken as a stress in the steel.
However, for the end sections near platforms where the pipeline may be free to move
there was concern that there would be tendency for the steel pipe to expand and move
relative to the concrete coating.

Differential temperature expansion occurs between the hot steel pipeline and the surrounding concrete weight coat because it is cooled by the surrounding seawater or loose back-fill. This differential expansion causes the enamel to be subject to shear forces along the length of the pipe similar to the "soil stress" phenomena on land between a pipeline and the surrounding trench back-fill. The magnitude of the shear stress in the enamel is limited by the restraint to longitudinal movement caused by friction between the concrete coating and the sea bed. A typical value of friction between concrete and sea bed would be between 0.15 and 0.20.

Over a period of time it was felt that substantial creep might occur within the coal tar enamel or at the interface with either the steel pipes or concrete. The movement so caused between concrete and steel may displace the cathodic protection anodes relative to the pipeline and could break the electrical connections between them thus destroying the cathodic protection. A limited research project was therefore initiated to examine the creep behaviour of coal tar enamels subject to shear, and to try to establish the probable magnitude of these effects.

4. EXPLORATORY WORK

In early 1977, B.P. in conjunction with Metrotect Ltd., set up a tank in which a number of short coated pipes were installed, mounted horizontally. No shear loads were applied but otherwise a positive attempt was made to make the test as realistic as possible. No physical measurements of movement were taken. Hot oil at temperature to 85°C was pumped through the steel pipes and the surrounding simulated seawater kept cool at approximately 4°C for a period of several months.

The samples were then removed from the tank and the concrete weight coating cut off. Visual and electrical holiday inspection of the coal tar enamel did not reveal any deterioration of the coating.

Following this work, Chevron U.K. Ltd., as operators for the Ninian Field, commissioned further work to evaluate these effects more fully and, if possible, to establish an acceptable limit of oil temperature for the grades of coal tar enamel currently in use in the North Sea.

In 1978 a new test tank was installed at Metrotect Ltd's laboratory in which four sample coated pipes could be mounted vertically. Provision was made for shearing the concrete relative to the steel pipe and measuring the resulting creep by means of dial gauges mounted between the steel pipe and the concrete surround. As in the previous tests hot oil was circulated through the pipe and provision made for cooling the surrounding sea water. Thermocouples installed in the samples enabled the temperature gradient through the composite to be monitored. The test rig is illustrated in Figure 1.

5. SAMPLE PREPARATION

The steel pipe samples comprised mild steel tube 1220mm long, 60mm diameter screwed at each end. These pipes were shotblasted by the method described in BS 4164:1967 and immediately coated with a synthetic primer based on chlorinated rubber (complying with Type B of BS 4164:1967 and AWWA C203-78) at the rate of 90 grams/square metre. The pipes were then flood coated with coal tar enamel (BS 4164:1967, grades 120/5 and 105/8). For all except two samples the enamel was reinforced with glass fibre inner wrap and a coal tar impregnated longitudinally reinforced glass fibre outer wrap spirally wound. The final enamel coating was 6mm (+ 1mm) in all cases. All coated pipes were then tested with a holiday detector set at 15 Kv.

Thermocouples were fixed (a) on the exterior pipe surface and (b) on the outer enamel surface at the centre of each pipe and at 100mm from each end. Concrete was then cast round the coated pipe by wet pouring in a rigid cylindrical PVC mould 200mm in diameter. The concrete mix was as follows.

Iron ore coarse aggregate (10mm)	38.2%
Sand aggregate	32.4%
Ordinary Portland cement (BS12)	23.4%

A water/cement ratio of approximately 0.42 was used to enable the concrete to be hand compacted. The concrete was cured at room temperature for 7 days and then, after removal of the PVC mould, maintained in a water bath at 27°C for a further 7 days minimum.

6. TESTING

Initially four test samples were made up but after a few weeks testing it became apparent that a full parametric study should be undertaken. While testing these first four samples numerous changes were made to the test rig and testing procedure to improve reliability and achieve the required temperature range. The parameters that we wished to study included:

a) The temperature of the oil.

b) The temperature gradient across the system.

c) The grade of coal tar enamel.

d) The effect of the glass fibre reinforcement in the coal tar enamel.

e) The level of stress.

f) The changes in the coal tar enamel under test conditions.

In each case we wished to measure creep with time. Funds, and more important, time only permitted the construction of a total of eleven coal tar enamel and concrete coated pipe samples.

It is immediately apparent that six independent parameters with at least two values of each parameter cannot be studied with eleven samples unless samples are re-used. Furthermore, it is desirable that tests be repeated using different samples to ensure repeatability of the results.

In practice these aims were not achieved and the testing programme had to be adjusted week by week as the work progressed depending upon the results obtained to date. This testing procedure therefore resulted in a series of continuous tests, three or four pipes at a time , in which the various parameters were changed at intervals. All temperatures and applied loads were recorded and a plot made of movement (creep) with time.

The tests were terminated when, for any particular sample, the aggregate of all movement to date reached the limit of the test apparatus or in the case of one or two samples when no movement could be measured almost irrespective of load.

7. TEMPERATURE

Oil temperature in the pipelines concerned are expected to be approximately 140 to 160°F (60 - 70°C) although it is recognised that this may rise to 170 - 190°F (77 - 88°C) on occasion for periods of a few weeks at times when cooler maintenance is being carried out. Oil velocities are such that turbulent flow occurs in the pipeline and it is assumed that the steel pipe will be at the same temperature as the oil. There is little field information concerning the interface temperatures between the coal tar enamel and the concrete. However, the sample proportions were selected so that temperature gradients would be similar across the coating to that expected in practice. Whilst attempts were made to get Reynolds number high enough to ensure turbulent flow in the sample, this was not found to be practicable for the size of tube and oil viscosities available. As a result laminar flow probably occured which caused a 10°C drop in the boundary layer between the flowing oil and the sample steel tube. We used the recorded thermocouple temperatures on the steel pipe as a true indication of temperature for prediction of creep behaviour from sample to prototype.

8. GRADE OF ENAMEL

High temperature enamel grade 120/5 has now become the standard enamel for North Sea application. As stated we wished to obtain a relationship between the maximum practicable oil temperature and the ring and ball softening point of the enamel. We therefore prepared samples using both 120/5 and 105/8 enamels. During the course of the testing it was noticed that the creep rates appeared to change with the age of the sample. Furthermore, on completion of each series of tests the concrete was again removed for inspection and remelt samples prepared for ring and ball and penetration tests. These samples invariably showed considerably higher results than the original enamel even though enamel stored at ambient conditions does not change in physical properties over this time period.

In a more recent series of tests specially prepared enamel samples in contact with concrete have been kept at elevated temperatures and tested at monthly intervals. These tests establish the above indications that the ring and ball softening point will increase with time under service temperatures. This effect will make a substantial difference to predicted behaviour of field applications particularly where unusually high temperatures will only occur for relatively short periods after a considerable time running at normal operating temperatures.

9. REINFORCEMENT

Considerable interest and speculation was aroused by the possible role of the fibre glass reinforcement on the enamel layer in the resisting of creep movement. For control purposes therefore two pipe samples, one of each grade of enamel, were constructed without reinforcement.

During test it was found that these samples did in fact creep more rapidly than their reinforced counterparts under comparable conditions and also that the creep followed a more predictable course. The reinforced specimens throughout the tests had shown a tendency for creep to occur at uneven rates and, particularly at the higher shear intensities, for sudden increases in rate to occur without apparent reason. We suggest that this occurs as a result of either disruption or realignment of the fibres. Some mechanical connections also probably occur since fibres will tend to connect hot areas of enamel adjacent to the pipe with relatively cool areas adjacent to the concrete. The enamel/concrete interface tends to be textured as a result of the pouring process and there is inevitably restraint to movement at this interface by mechanical interlock. Substantial movements therefore, when they were forced to occur, tended to occur as a shearing action at the junction between the primer and enamel as would be expected since this is not only the hottest zone of the enamel but is also a smooth interface.

CONCLUSIONS

1. Grades of Enamel

Pipeline enamels are produced to specifications which cover a range of softening
point (R & B) and penetration characteristics. These properties vary from high
SP (R & B) and low penetration, made for hot line applications to the much lower
SP (R & B) medium penetration (softer) enamels, for use in temperate or colder
climates.

A firm conclusion which can be drawn from this investigation is the importance of
selecting the correct grade of coal tar enamel. Figure 2 shows results with 105/8
and 120/5 grades of enamel. After 21 days the creep with the softer 105/8 amounts
to 60mm and is still steadily increasing. Although the 120/5 grade enamel was
heated to a significantly higher temperature, the creep was less than 10mm after
25 days and appeared to be levelling off. This was confirmed by running the test
for a further 7 days.

At the creep rate experienced in our tests it is concluded that 105/8 grade enamel is
an unsuitable choice for the pipelines planned at operational temperatures of 95 $^{\circ}$C
on the pipe wall.

2. Effects of Glass Reinforcement

A comparison of the creep behaviour of 105/8 grade enamel with and without the
glass reinforcement is illustrated in Figure 3. Without reinforcement the creep was
at a steady rate from the time of application of the stress to the heated pipe. After
3 days the rate was increased by the doubling of the applied stress to a new fairly
steady level. The effect of the reinforcement in sample 'A' has been to inhibit the
creep during the first two weeks, after which it proceeded at a rate similar to that
of the sample 'C' without reinforcement.

Of the eight pipes protected with 120/5 grade enamel and glass reinforcement the
behaviour has been very variable. In 3 samples including 'B' shown in Figure 3,
the reinforcement had no apparent effect but it was not possible to induce
appreciable creep in the remainder at 90 - 100 $^{\circ}$C under a stress within the range of
interest. It was, however, possible to overcome the resistance to creep in two
instances by increasing the stress to 8 - 9 kN/m^2.

It is concluded that unavoidable inconsistences in wrapping the pipes may result in
a marked resistance to creep, which would be beneficial in practice, but this is an
effect which would be difficult to exploit to practical advantage.

3. Temperature

One pipe coated with 120/5 grade enamel, but without reinforcement, was subjected
to successively increasing temperature, the results being shown in Figure 4.
The most significant feature is that at each temperature, at least up to 95°C, there
is a slowing up of the creep after a week or two. It has been found that the heating
of the enamel at these high temperatures, causes within the first few weeks an
increase in softening point, due presumably to a loss of the most volatile of the tar
components into the concrete. The effect of this appears to be to retard the rate of
creep to an acceptable level. 90 days heating at temperatures of 72 - 110°C
resulted in barely 3mm creep. Further tests are required to define more precisely
the maximum creep which may occur with 120/5 grade enamel at given temperature
and stress conditions, but there is ground for optimism that the suitability of this
grade of enamel for use under North Sea conditions can be established.

4. Primer Properties

Pipeline primers are now practically all synthetic, fast drying primers based upon chlorinated rubber (C.R.) or similar polymers in fast drying solvents. They are much more reliable in performance than the early "cut back" primers and in addition permit faster coating of the pipe.

In a programme of testing using Scanning Electron Microscopy (S.E.M.) it has been established that in correctly formulated, adequately stabilised, synthetic primers there are no deleterious effects on the primer/enamel, nor primer/steel bond after heating the composite system at 100°C. It is stressed however, that the correct formulation and stabilisation of the system is of paramount importance in the production of primers for use at high "in service" temperatures.

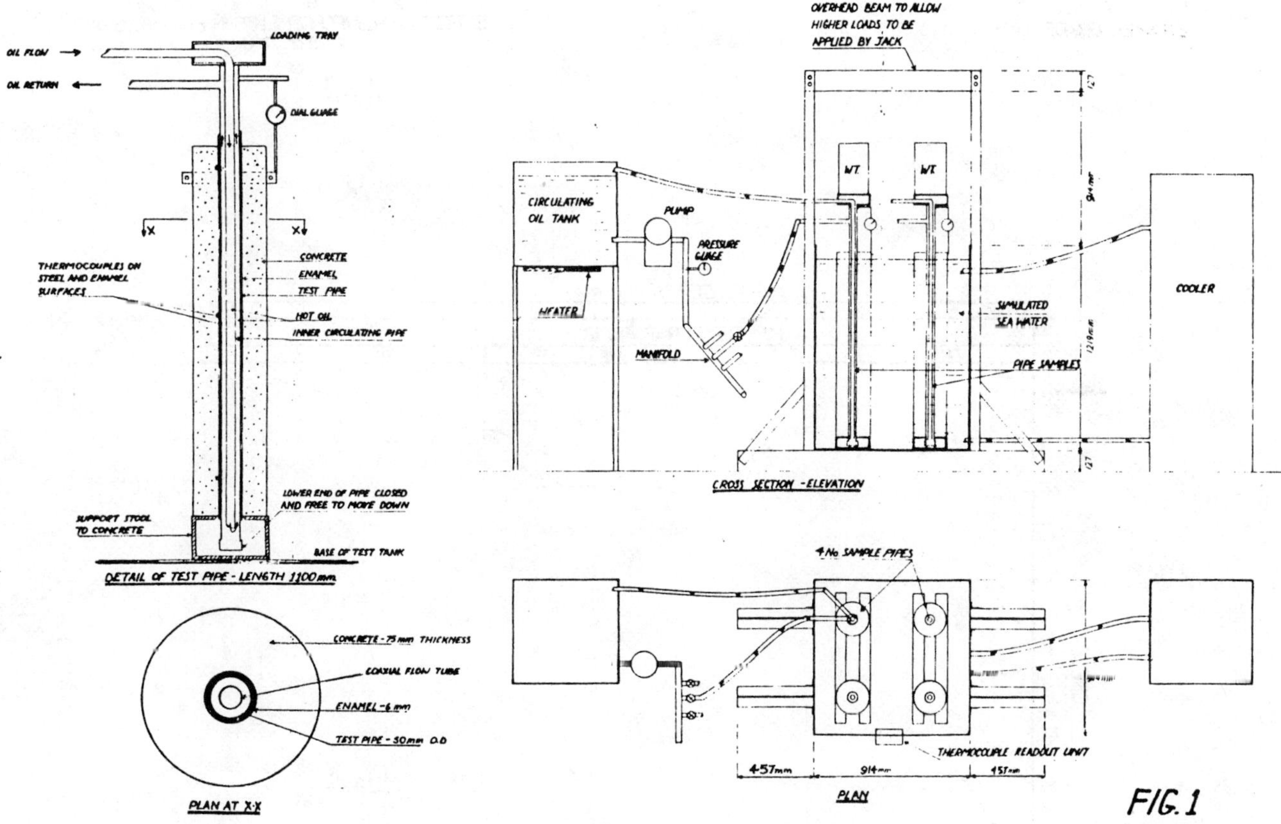

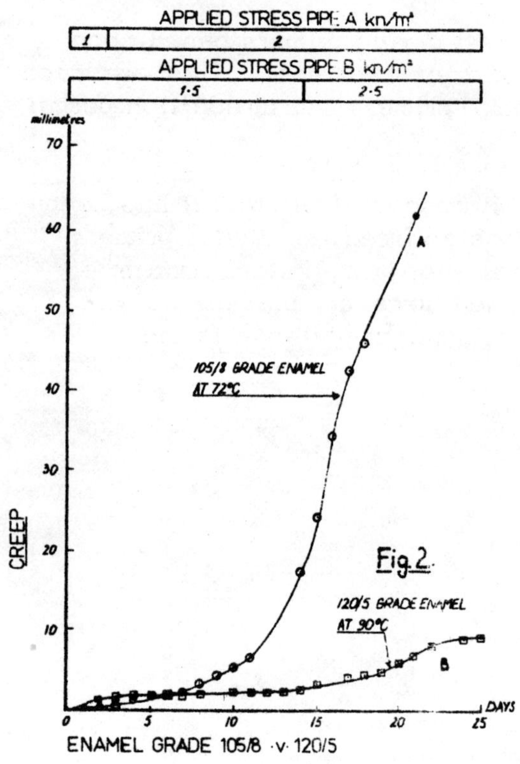

ENAMEL GRADE 105/8 ·v· 120/5

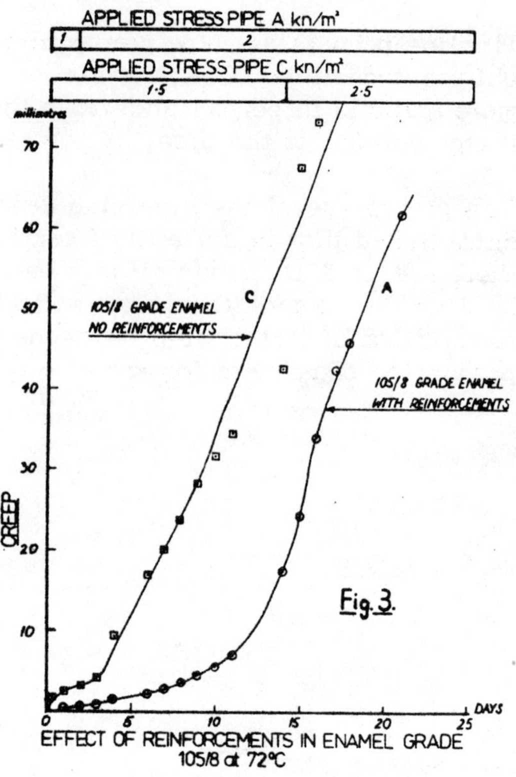

EFFECT OF REINFORCEMENTS IN ENAMEL GRADE
105/8 at 72°C

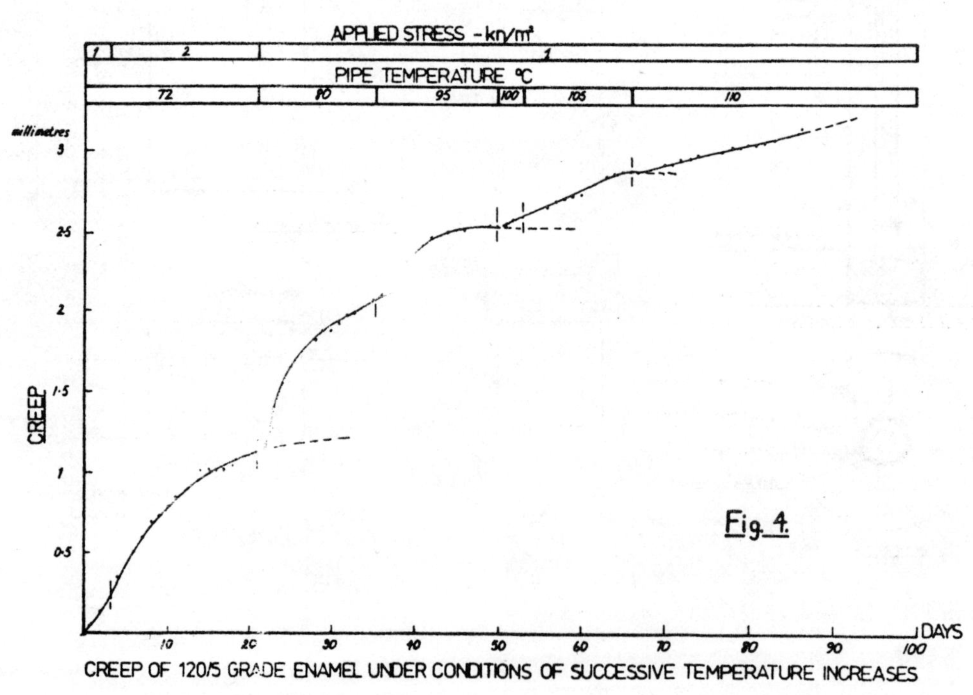

CREEP OF 120/5 GRADE ENAMEL UNDER CONDITIONS OF SUCCESSIVE TEMPERATURE INCREASES

internal and external
protection of pipes

QUALITY TESTS OF PLASTIC COATED PIPES FOR THE OIL AND GAS INDUSTRY AND THE QUALITY CONTROL FOR PLASTIC POWDER FOR COATING PROCESSES

M.H. Akstinat, Dr.-Ing., Dipl.-Chem, and W. Schuhbauer, Dipl.-Chem.

Technical University of Clausthal, Federal Republic of Germany

Summary

The use of plastic coated pipes in the oil and gas industry has increased considerably during recent years, whereby substantial economic advantages have been particularly important.

In comparison with conventional coatings stringent quality requirements are imposed on plastic coated pipes for use in petroleum and natural gas exploration or production, in order to fulfil the extreme demands in the oil and gas industry. At the Institute of Petroleum Engineering and Drilling Technology (Federal Republic of Germany) a program was therefore set up for testing the quality of plastic coated pipes. This includes the most important tests such as:

> measurement of coating thickness, curing,
> porosity, adhesion, elasticity, abrasion resistance,
> impact resistance, torsional strength, tensile strength,
> autoclave tests in NaC1 solutions and in drilling muds at
> various pH-values, acid resistance, resistance to water
> vapour, and surface roughness.

The individual procedures and instruments are described and their significance is assessed. Results of the quality tests performed on typical pipe segments from different producers are mentioned.

Since the quality of a plastic coating also depends particularly on the properties of the original plastic powder, a procedure was likewise developed for testing the plastic powder, in order to assure a consistent quality of the coatings. This testing procedure for the plastic powder is also described.

Held at Imperial College, London, England.
Organised and Sponsored by BHRA Fluid Engineering

1 Properties and functions of plastic coated pipes in the petroleum and natural gas industry

During the past two decades duroplastic coatings have gained particular importance in the field of passive corrosion protection of metals /17/. Because of the very specific technical requirements, the skills and experience involved in the coating technology, a new, rapidly progressing and developing sector has arisen in this area of corrosion protection.

Entirely new types of corrosion protection systems have been developed, new procedures have been formulated and wide ranges of application have become feasible. The petroleum and natural gas industry, which suffers yearly corrosion damages ranging in the millions, has recognized the value of such protective measures against corrosion and is employing to an increasing extent plastic coated drill pipes and well tubing /18, 20/.

The extreme conditions to which the plastic coated pipes are exposed in the oil and gas industry are of numerous kinds /21, 22/:

- mechanical effects (shock and impact; elastic deformations; overpressure; solids, liquids and gases in motion)

- thermal effects (high temperatures; attack, by hot liquids and vapours; prolonged, brief, or intermittent thermal stresses)

- chemical effects (acids; bases; salt solutions; vapours; corrosive gases; reservoir water, etc. /16, 18, 23/).

Under consideration that no universal construction material exists it cannot be expected that duroplastic coatings can solve all corrosion problems of the oil and gas industry. The development of new plastic coatings is, however, progressing rapidly at present, and it can be anticipated that substantial improvements in coating technology /21, 22/ and in the composition of plastic powders can be achieved during the coming years.

The scope of this report is restricted exclusively to those coating systems whose organic binder (in combination with inert inorganic fillers) yield hard, infusible and insoluble protective duroplastic coatings after sintering on a preheated substrate (pipe surface) and subsequent fusion and curing. The curing of the plastic coatings considered here usually is carried out both by thermal and by chemical catalytic polymerization. This category includes especially phenolic, epoxy /16/ and silicone resins. Chemical catalytically cured duroplastic coatings based on polyurethanes (curing with polyisocyanates) are also gaining increased significance.

Besides corrosion protection the plastic coating of pipes can enhance the transport efficiency by 5 to 10 %. Today in pipeline construction with large nominal sizes the use of internally coated pipes is therefore regarded worldwide as the state of the art /19/. For this application a maximal roughness of 6 to 8 μm is desirable.

In addition to corrosion protection and improvement of the transport efficiency a plastic coating also is effective in suppressing paraffin deposition in oil and gas wells /9, 19, 23 - 28/. JORDA /9/ demonstrated that the quantity and type of paraffin deposition in oil wells can be reduced considerably by a reduction of the surface roughness. However, wetting phenomena between the plastic coating and the paraffins probably exert some further influence on the precipitation rate of paraffins.

2 Conditions and requirements on the surface properties of pipes for the application of protective plastic coatings /10/

A plastic coating can fulfill its purpose of corrosion protection only, if it adheres completely to the surface of the pipe material and totally isolates the coated substrate from the corrosive medium.

Efforts to fulfill these requirements encounter in part considerable difficulties because of the prescribed design of the pipes (for example connections and threads) and to a large extent because of the coating technology.

Sharp edges, points, burrs and perforations can lead to critical reductions in the coating thickness .   Nonuniform preheating or deviations from the optimal bonding temperature also result in defective coatings (compare VDI recommendation 2532).

Furthermore, the pipe surface to be coated should not be already corroded as a result
of prolonged storage in the open; in any case the surface must be smooth, dense and
free of pores, laminations, blisters, drawing and rolling defects, etc.

In the case of already used pipes,which are to be coated for the first time or
recoated,the surface is usually more or less severely corroded and is suitable for
coating only under special circumstances and after intensive pretreatment. Corroded
regions, pits, grooves and spongy areas are frequently causes of coating defects and
thus starting points of corrosion during use in the bore hole.

The quality and working life of a pipe coating depends to a large extent on the
correct preparation of the pipe surface to be protected. In order to ensure optimal
bonding between the plastic coating and the metallic surface the latter must be
carefully cleaned and dried prior to coating. Moreover improved adhesion of the
coating can be achieved by means of roughening the base material. An appropriate
pretreatment process consists in dry sandblasting, whereby the degree of rust removal
of 3 (bright metal) is to be attained.

It is important that the metallic surfaces are coated immediately following the
pretreatment since otherwise a film of rust or an imperceptible water film can form
on the active surface as a result of the condensation of moisture; this impairs
adhesion. As an alternative measure of temporary protection a primer or a phosphatizing
process can be employed.

3 Quality tests on plastic coatings for use in the oil and gas industry

Plastic coated pipes are exposed to extreme conditions in the petroleum and natural
gas industry. In order to fulfill the rigorous requirements in this field of application both manufacturers /2, 20/ and users /3/ have usually developed their own
procedures for quality testing, for want of generally accepted testing standards.
Because of the increasing demand for suitable testing procedures for plastic coated
pipes NACE /11, 22/ has also occupied itself with the development of standardized
testing methods, which have originated in various fields of endeavor. However,
generally accepted testing procedures for plastic coated pipes do not yet exist. It
is therefore the purpose of this article to summarize the hitherto applied quality
tests such as

- visual inspection
- coating thickness
- porosity
- curing
- adhesion
- abrasion
- impact

- tensile stress
- torsional stress
- autoclave test
- acid resistance
- resistance to water vapour
- surface roughness

and to examine their reliability and conclusiveness as applied to real coating systems.

3.1 Visual inspection

The visual inspection of plastic coated surfaces is an important preliminary testing
criteria and already allows extensive statements about the quality of coatings.
Geometrical defects such as spiral grooves or striations indicate faulty operation
of the coating equipment,e.g.. An "orange peel effect" usually has its origin in a
nonuniform powder application or in deviations from the optimum coating temperature.
A so-called "snot-nose effect" is generally caused by incorrectly adjusted viscosity
values during the fusing process. Further serious coating defects are, among others,
occlusions of metal filings, incompletely cured polymer powder, as well as cracks
and coarse pores, which already can be detected by means of a visual examination
(Fig. 1 and 2). The majority of the faults present in plastic coatings usually can
be recognized by the visual inspection. Hence this primary check should precede all
further tests.

3.2 Measurement of the coating thickness

Among other factors the maintaining of a uniform coating thickness vitally affects
the quality of a good coating. A good coating should show no substantial positive
or negative deviations from the required thickness at all points.

In general the measurement of the coating thickness involves a nondestructive test,

which should be performed by means of an instrument adapted to the thickness range
in question.

The measurements discussed were performed with an apparatus called MINITECTOR 150 FN
(ELCOMETER INSTRUMENTS LTD.). The coating thickness is hereby measured by means of
electromagnetic and eddy current sensing coils, which are housed in a measuring
probe.

The coating thickness should be measured at several points statistically distributed
over the surface of the specimen to be tested in order to obtain representative
values for the coating thickness (mean values). The scatter of individual measured
values should thereby not exceed $\pm$ 10 % of the average thickness of the coating /22/.
A coating thickness of 150 $\pm$ 50 μm was found to be optimal for drill pipes and
tubing coated with epoxy resin.

Typical average values for the thickness of epoxy resin coatings from production are
presented in table 1.

3.3 <u>Porosity test /1, 22, 23/</u>

The demand for a 100 % nonporous coating results from the goal of completely isolating
the material to be protected from a corrosive medium.

The coating technology is relatively complex, because the applied powder coatings
are subject to physical as well as chemical processes prior to final curing.
Processes such as the separation of reaction products, the volatilization of solvents,
occlusion of foreign particles or shrinkage effects can generate faults in an
otherwise homogeneous coating (Fig. 3). Directives for producing a nonporous coating
are given in the VDI recommendation No. 2532. In general, any coating defect, through
which a corrosive medium can come into contact with the material to be protected, is
defined as a holiday /8/. This definition includes microscopically small cracks or
pinholes, which can arise during coating, as well as the occlusion of metal filings
resulting from faulty or insufficient preparation of the surface to be coated.
Mechanical damage, as may occur during field operations, may also produce holidays. In
practice all defects, which exhibit a lower electrical resistance than a prescribed
limiting value, are designated as holiday. Very often a resistance of 80 000 ohms
is taken as limiting value /22/.

The principle of the porosity test is based on a resistance measurement. The tests
discussed were performed according to the direct-current electrolytic wet-test
method with a sponge electrode and an acoustic pore-detecting device (PORTECTOR /
Model 169). The no-load voltage of the instrument was 50 - 100 V, the measuring
voltage was 90 V and the testing speed of the sponge electrode amounted to 0,3 m$\cdot$s^{-1}.
The electrolyte used was prepared with the following volume contents:

    94 % dist. water
     1 % Ammonia solution 25 %
     5 % Ethanol 90 %

The following points must be considered while the measurements are being performed:

- Adaptation of the voltage to the coating thickness
- Avoidance of excessive quantities of liquids
- Adaptation of the contact area of the sponge electrode

If the voltage is chosen too low for the coating thickness present an erroneous
indication that holidays are absent may result. Excessive moistening of the coating
may also lead to wrong values due to pores farther distant because of bridging
contact. In the case of excessively large sponge contact area several small pores
with insufficient individual conductance may be contacted simultaneously, thus
resulting in one indication,which would otherwise not occur.

3.4 <u>Curing test</u>

A test, which is important for the coating manufacturer, involves the stability of
a protective plastic coating toward solvents. If,e. g.,a plastic coating is not
completely cured, an increased sensitivity toward polar solvents becomes evident
in this test. In the case of insufficiently cured coatings deterioration phenomena
of the coating may be observed during storage in solvents, these results in peeling,
flaking, swelling, softening, shrinkage, tearing, blistering, etc. The investigations

discussed here were conducted by means of a 30 minute exposure to Methylethyl-
ketone (MEK) or Acetone. For the purpose of judging the degree of curing on typical
specimens the pencil hardness test was performed on the sample surfaces still moist
with solvent.

The pencil hardness test applied provides a rapid classification in 14 degress of
hardness, which are denoted as follows:

Degrees of pencil hardness

6 B   5 B   4 B   3 B   2 B   B   HB   H   2 H   3 H   4 H   5 H   6 H   7 H
very soft ⟶ ⟶ soft ⟶ ⟶ normal   ⟶ ⟶ hard   ⟶ ⟶ very hard

Typical test results are presented in table 2. In the present experiments it became
obvious, that tests conducted with MEK lead to a better differentiation about the
resistance to solvents, in comparison to those with Acetone.

3.5 Adhesion

The evaluation of the bonding strength of a plastic coating to the base material
represents another important functional test, which may be performed by any of the
following methods.

3.5.1 Cross-cut test (according to DIN 53 151)

The cross-cut test provides information about the adhesion between the coating and
the base material. Moreover it allows, to a certain extent, conclusions about further
properties such as hardness and elasticity of the coating layer. Single or multicutter
instruments with various knife spacings are available for the testing of plastic
coatings. The choice of the appropriate cutting device is fixed by the range of
coating thickness in question. The sample plates investigated here were tested with
a multicutter device (Fig. 4), whereas pipe sections were tested with a single cutter
device.

However, the rating, which was carried out according to DIN 53 121 /13/ by use of
the evaluation table shown in figure 5, does not allow sufficient differentiation
for a practical classification of the coating quality.

3.5.2 Measurement of bonding strength (based on DIN 53 232 and DIN 51 221)

In order to confirm the results obtained by the cross-cut test and to avoid any
erroneous conclusions a further adhesion test of a coating should always be conducted.
The physically best defined test procedure is the vertical pulling off of a coating
from a base material (DIN 53 252 /15/) (Fig. 6). However, a prerequisite for this
is the availability of suitable adhesives, which possess sufficient bonding strength
and do not alter the properties of a plastic coating. For performing this test
special test plugs are fastened to the coating by means of an appropriate adhesive
(e. g. Araldite, Sicomet 99). Once the adhesive has thoroughly cured the test plug
is pulled off with an adhesive strength tester. The pulling force per unit of area
is recorded in $N \cdot mm^{-2}$. In addition a visual inspection of the site of breaking
is necessary in order to permit comparative conclusions.

The present measurements were performed with the adhesive strength tester ELCOMETER /
Model 106 (ERICHSEN) on various plastic coatings, whereby the bonding strength was
registered by means of a maximum indicator pointer. Typical values of the bonding
strength are presented in table 3.

3.6 Abrasion resistance (Falling-Sand Method) /6/

During field operations a coating is usually subjected to considerable mechanical
stresses, which can vary both in type and intensity. This complicated attack on a
plastic coating, as it occurs in practice, is extremely difficult to simulate in a
laboratory test. A simple test method, which reproduces the effects occurring in
practice, is provided by the erosive exposure to falling sand. The tests are conducted
according to ASTM Standard D 968 /12/ and NACE-Standard TM-03-75 /11/. The abrasion-
testing apparatus acc. to STOCK (Fig. 7) (ERICHSEN) as employed here consists
essentially of a vertical fall tube through which sand impinges upon a test
specimen inclined at an angle of 45° /6/.

Since the specimens tested here vary greatly in coating thickness the evaluation procedure was modified. As a measure for the abrasion resistance of a coated surface an abrasion coefficient Ψ is defined as follows:

$$\text{Abrasion coefficient } \Psi = \frac{\text{Loss in mass of coated surface in mg}}{100 \text{ l sharp-edged Quartz sand}}$$

(Typical values of the abrasion coefficient for flat sheet specimens and pipe segments in table 4) The generally more severe abrasion in the case of pipe segments, as compared to flat samples, should be viewed as a particularly significant result, that is, the shape of the surface exposed to the falling sand is also decisive.

3.7 Impact test (acc. to ASTM D 2794-69) /14/

The handling of coated pipes and tubing in the oil and gas industry usually proceeds under uncontrolled, rough conditions, whereby shocks and blows of various kinds cannot be avoided. An impact can be implemented either directly or indirectly (from the opposite side of the coated surface) by blunt or sharp objects; indirect impacts usually causes less damage. Good plastic coatings should withstand small shocks and blows from blunt objects and remain fully intact; in case of severe blows from blunt or sharp objects only slight damages to the plastic coatings should occur.

The oil and gas industry in general demands both direct and indirect shock and impact testing of coated materials.

The procedure for direct impact testing is based on ASTM D 2794-69; a weight of 10 N with a hardened spherical head is dropped successively from heights of 100, 180 and 200 cm onto a coated surface.

The evaluation of the damages to the plastic coating is conducted visually and allows a qualitative classification of different coatings. Typical results of such an evaluation are summarized in table 5. In the case of a faulty, brittle coating (Fig. 8) not only the region immediately subjected to the impact is destroyed (inner compacted circle); usually the coating also chips off within a larger area. Such plastic coatings are unsuitable for protecting tubing and drill pipes.

Subjecting of the coated surfaces to reverse impact is likewise accomplished according to ASTM D 2794-69 with 20 N·m because of the considerable thickness (usually > 5 mm). The inspection of the plastic coating is performed visually here, too. For all investigations hitherto conducted no visible damage to the plastic coating could be detected during the reverse impact test, even with an impact of >20 N·m.

Since the reproducibility of the test results is not very good at least 3 parallel tests should always be performed.

3.8 Tensile strength and breaking test

Drill pipes and tubing are subjected to high tensile stresses during service. All of the tensile stresses exerted on the metal pipes within the elastic region must also be withstood by the adhering plastic coating without tearing or separating from the metallic surface. Repeated cycles of stressing and destressing (during installation and pulling of the drill string) must likewise be withstood by suitable plastic coatings without damage.

For this pupose conventional tensile tests on plastic coated pipe segments are especially appropriate; in the elastic range of the base material stressing and destressing can hereby be repeated several times. The samples to be tested are prepared according to DIN 50 125 and DIN 50 140 /29, 30/ (Fig. 9).

In order to register damages and movements of the plastic coating within the elastic deformation range of the steel, wire strain gauges were initially glued to the coated surfaces. However, numerous investigations have shown that a qualitative rating of individual products with the help of fine wire strain gauges is impossible. It was demonstrated that most of the plastic coatings tested here follow all the motions of the base metal within the elastic range and do not exhibit any visually recognizable damage. Very brittle and inelastic coatings,

such as cannot be employed for coating of oil-field pipes, usually show transverse
cracks (90° to the direction of tension) already in the elastic range of the base
metal (stress up to 80 % of the elastic limit) (Fig. 10).

A differentiated evaluation of good to very good plastic coatings in the elastic range
of the base metal is, however, no longer possible.

For this reason the tensile test on plastic coated pipe segments was supplemented by
a breaking test.

It could be shown, that although considerably higher stresses are imposed on plastic
coatings than in practical field service when the elastic range of the base metal is
exceeded, a good qualitative classification of various coatings is thus made
possible (Table 6).

The breaking test on coated pipes – particularly with the evaluation in the
constriction range – provides interesting and conclusive results concerning adhesion,
elasticity and temperature stability (considerable heat is evolved in the constriction
range) (Fig. 11 and 12).

3.9 Torsional test

During field operations drill pipes are subjected to considerable torsional stress.
Properly cured plastic coatings must withstand any twisting to the same extent as
the base material without loss of adhesion.

For a torsional test a narrow strip cut from the plastic coated pipe sample to be
investigated (width = double the wall thickness) is securely clamped at one end.
The other end of the strip is rotated until a twist angle of 180° is attained
(Fig. 13). The distance between the clamped end and the point of application of the
twisting force is equal to 25 cm.

The tests provide information about the elasticity of a coating, whereby possible
aging effects must be considered, too (a second examination of the specimen is
performed after a 7-day storage at room temperature). Typical test results for
epoxy-powder coatings of various origin are presented in table 7. Inferior polymer
qualities indicated by the torsional test also correspond to unsatisfactory results
in the breaking test (compare table 6).

3.10 Autoclave test

Of particular importance for plastic coatings of drill pipe is, e. g., their
stability toward drilling muds at various pH-values and temperatures. Hence,
appropriate tests are essential for ascertaining the limits of applicability for
various types of polymers and for optimizing novel polymer mixtures.

Typical water-based drilling muds are employed in the pH-range >7, while the upper
limit for the working temperature is currently regarded to be approximately 150 °C.
On the basis of these considerations

- NaCl-solutions and
- typical drilling muds

with various pH-values and at different temperatures are therefore chosen for testing
plastic coatings. A test duration of 100 h is regarded as sufficient for assessing
the quality of a polymer coating. For determining the loss of adhesion of plastic
coatings a cross-cut test is performed acc. to DIN 53 151 after the autoclave
treatment /13/.

3.10.1 Tests with NaCl-solutions

In the present test series a 3 % NaCl solution was employed as base solution, the
desired pH-value can be adjusted to conform to the particular requirements.
Endeavors were made to attain all desired pH-values up to pH 12 at the corresponding
test temperatures. However this was possible only up to pH 10 at 150 °C, a further
increase in the pH-value could not be realized at 150 °C with the preselected medium.

Hence, the following solutions were employed for performing the autoclave tests
(Table 8 and Figure 14):

A 1, B 1, C 1   = 30 g NaCl

                         + 2 ml 0,1 n HCl   in 1 l dist. water

A 3, B 3, C 3   = 30 g NaCl

                         + 0,1 g CaO   in 1 l dist. water

A 2   = 30 g NaCl

                         + 15 g CaO   in 1 l dist. water

B 2, C 2   = 30 g NaCl

                         + 12 g CaO

                         + 25 g NaOH   in 1 l dist. water

The initial pH-values and the changes in the pH-values at the corresponding test temperatures of 120, 135 and 150 °C are listed in table 8. It was attempted to approximate the desired pH-values as closely as possible with the least possible number of solutions.

In order to obtain reliable results from autoclave tests it is necessary to take representative samples or to test a sufficient number of samples in parallel.

After the autoclave tests in the pH-range from 7 to 10 at temperatures of 120 to 150°C and p = 160 bar each sample is inspected visually and subjected to a cross-cut test acc. to DIN (results in Table 9).

From the tests with 3 % NaCl-solution (100 h) at various pH-values it can be inferred that

- with increasing pH-value and rising temperature a definite qualitative
  classification of the various coatings is possible and that

- unexpected scatter for individual specimens can lead to wrong conclusions
  unless a sufficient number of specimens are tested and judged as a whole.

3.10.2 Tests with standard drilling mud

For comparing with a 3 % NaCl-solution the plastic coatings were also exposed to a drilling mud with a high alkaline earth content (without weighting agent Baryte) at various pH-values and temperatures (Table 8 and Fig. 15).

Composition of drilling mud:

              30 g NaCl

        +   2 g CaO

        +   400 g $MgCl_2$ · $6H_2O$

        +   50 g $CaCl_2$ · $6H_2O$

        +   30 g Magne-Magic (OIL BASE/Germany)

        +   30 g Viscogel 618 (SCHOLTEN / Holland)

        +   x g NaOH in y g H_2O predissolved

                    in 1 l dist. water

for solutions   A 7, B 7, C 7   x = 0   y = 0
                 A 4, B 4, C 4   x = 126   y = 160
                 A 5, B 5, C 5   x = 127   y = 160
                 A 6, B 6, C 6   x = 200   y = 200

Despite of the fact that extensive precipitation occurs at pH-values ⩾ 8,0 such suspensions were also used as test media.

In order to achieve meaningful results a representative number of samples had to be tested here, too.

After the autoclave test with the standard drilling mud each specimen is individually inspected and subjected to a cross-cut test acc. to DIN 53 151   (see Table 10).

In comparison to the tests with the 3 % NaCl-solution the following was observed for the standard drilling mud:

- At a pH >10 severe damage occurs on nearly all coatings, whereas in the pH-range <10 no significant quality loss occurs.

- The tests in drilling mud systems did not provide any further qualitative classifications, which are essentially superior to that with the 3 % NaCl-solution.

3.11 Acid resistance

In oil and gas fields acidic corrosive media frequently occur (e. g. sour gas), furthermore highly concentrated acids are often used for well treatment (acidizing); plastic coated pipes must therefore be resistant to such agressive media.

For this reason it is necessary to test coatings intended for the protection of pipes for the oil and gas industry for their acid resistance.

The acid resistance test on a plastic coating is generally carried out acc. to DIN 51 167 with 15 % HCl (+ 0,1 % inhibitor) at 80 °C and at boiling temperature /7/.

The coatings are hereby exposed to both the liquid and vapour phases for 1 h (test apparatus in Fig. 16).

No loss of adhesion (cross-cut test) or blistering could be observed with any of the coatings tested here; either exfoliation occurred (coating fully removed from base material) or there was no change at all. Hence, the pencil hardness test is recommended as a supplementary test method for samples,which show no evidence of damage (results in Table 11).

The following conclusions are possible on the basis of the present test results:

- in general the hardness of plastic coatings is decreased by an acid treatment
- porous coatings are usually completely removed from the base metal

The coatings of some competitors passed the acid resistance test without any particular deterioration. However, it must be emphasized, that the results of this test strongly depend on the porosity of the individual specimens.

3.12 Resistance to water vapour

An important selection criteria for plastic coatings is their resistance to water vapour, which occurs during both crude oil and natural gas production.

The testing of plastic coatings with respect to water vapour has hitherto been conducted acc. to DIN 51 165 /7/. In this test coated specimens are exposed to the gas phase for 60 min (boiling water) and subsequently inspected visually.

Damage to the coatings (blistering) or inferior results of the cross-cut test could not hitherto be observed with this test method.

As a further test possibility following the water vapour treatment the pencil-hardness test is also suggested. However, only slight losses of hardness can hereby be detected (Table 12).

A differentiated qualitative classification is hardly feasible by means of this test method.

More extensive considerations of this test procedure led to a modified technique, in which the damage to plastic coatings is investigated by means of cavitation. By this method the formation of cavities (blisters) due to degassing or vaporization in streaming liquids as a result of a sudden pressure drop should be investigated. However, results from this test method are not yet available.

3.13 Determination of the surface roughness

The surface roughness of a coated pipe is governed on one hand by the surface quality of the base metal and its pretreatment, and on the other hand by the fusion process of the polymer powder.

Among other parameters the surface roughness of the base metal strongly affects the consumption of polymer powder (see also DIN 53 220) and the uniformity of the resulting plastic coatings. However, after sandblasting the consumption of polymer powder in the case of corroded steel surfaces will differ greatly from that for

noncorroded, smooth new surfaces; that is, for old pipes to be coated or recoated
a higher degree of wrinkling and porosity must be expected. For protective duroplastic
coatings with a thickness of about 200 µm an average rugosity of the base metal of
50 to 60 µm has proven to be especially advantageous.
In the case of faulty fusing behavior (insufficient viscosity of the polymer melt,
poor wetting etc.) of the polymer powder inferior surface qualities are likewise to
be expected.

By means of roughness profiles taken from coated samples "smooth" surfaces can be
clearly distinguished from "rough" ones, whereby the average rugosity of a good
coating should lie below 5 µm.

A low degree of roughness for plastic coatings influences in particular undesirable
deposits on the pipe wall and also reduces friction. The "smoother" a surface is,
the lower its frictional resistance and tendency for paraffin deposition will be;
this implies higher rates of production and lower servicing costs.

The determination of the surface reoughness can be performed by means of a PERTH-O-
METER. Typical roughness profiles are shown in Fig. 17 and 18.

The investigation of various epoxy-resin coatings on pipe segments yielded clearly
differentiated conclusions about the surface quality (Table 13).

4 Assessment of the conclusiveness of the individual test methods

Under consideration of numerous test series for evaluation plastic coated pipes for
the oil and gas industry an appraisal of the test methods described here should be
made. Where applicable alternate testing methods should be indicated.

4.1 Visual inspection:

Important and meaningful preliminary test; provides information about faulty
coating technology

4.2 Measurement of coating thickness:

Necessary quality test for ascertaining minimal and maximal coating thicknesses;
if these limits are not attained or exceeded, definite reduction in quality
can be expected

4.3 Porosity test:

Important testing method; high degree of porosity leads to premature attack
by corrosion /10/; present sponge electrode method /11/ allows only coarse
classification of porosity; improvement of procedure for differentiation of
macro- and micropores is required

4.4 Curing:

Nonspecific and hardly differentiating test for ascertaining the degree of
curing of a plastic coating for the manufacturer; improvement of the method
necessary

4.5 Adhesion:

Both the cross-cut test /13/ and the removal by means of a pull-off device are
not very conclusive for judging the bonding strength of plastic coatings;
improvements of the methods are necessary

4.6 Abrasion resistance:

Results of the abrasion test with falling sand do not allow clear differentiation
between various coatings and are not conclusive; a substantial improvement of the
testing method would be possible by the standardized use of a TABER-Abraser

4.7 Impact test:

Important, but hardly differentiating method; does permit experienced testers to
get further conclusions about elasticity and adhesion of a plastic coating

4.8 Tensile strength and breaking test:

Tensile strength test is conclusive only for qualitatively unsuitable, brittle
coatings/differentiating conclusions about the quality of good to excellent
coatings possible only by breaking test; important and necessary test procedure

4.9 Torsional test:

Test method conclusive only for poor, brittle coatings; similar information can
also be obtained from tensile strength and breaking test

4.10 Autoclave test: Important and necessary test method; results with drilling muds
provide no more information than those with 3 % NaCl-solution; hence the latter
alone suffice fully

4.11 Acid resistance:

Hardly conclusive and poorly differentiating test procedure; improvement is
necessary

4.12 Water vapour test:

Hardly conclusive and poorly differentiating test procedure; improvement is
required; future extension of the test to water vapour diffusion is desirable

4.13 Surface roughness:

Important and necessary test, expecially for well tubing; roughness profile
permits useful conclusions for quality classification; future extension of the
test to include paraffin deposition rate is desirable

5 Investigation of coating powders

The production of superior plastic coatings depends on the one hand on the coating
technology and on the other on the constancy of the quality of the powders used for
the coatings. Hence, the standardization of the coating technology and of the polymer
powder is of primary importance.

Quality control processes for polymer powders were developed at the Institute of
Petroleum Engineering (Clausthal), these include the investigation of the particle-
size distribution and of the fusing behavior of the polymer powders.

5.1 Particle-size distribution

The grain size and form (Fig. 19 / DIN 53 206) of plastic particles are decisive for
the thickness, uniformity and melting behavior of a plastic coating. For this reason
the particle-size distribution of plastic powders are of vital importance.

As an aid for exactly determining a sieve-analysis curve or for investigating the
particle-size distribution, an air-draught sieve is appropriate (Fig. 20); this has
often proven useful, especially for fine powders (see DIN 53 734). The conventional
mechanical sieve-analysis with standardized sets of sieves is often susceptible
to error, because polymer powders ordinarily contain a high proportion of fine particles, which tend to form aggregates and agglomerates and thus cannot pass through
the corresponding sieves without difficulties. A further drawback of the conventional
sieve-analysis is that the sieve column usually does not close air tight and thus
that fine polymer dust can easily be transported away uncontrollably. Sifting with
the help of brushes is not recommended, since polymer powders frequently become
electrostatically charged and consequently cannot easily be retained on the sieve.

The Grindometer test acc. to HEGMAN (Fig. 21 / DIN 53 203) is suitable as a rapid
method for ascertaining the limiting values for the maximal grain size of plastic
powders. The plastic powder to be tested is thereby dispersed in an inert oil and subjected to the test.

5.2 Fusing behavior

A further important criterion for selecting polymer powders is their behavior during
fusion. A test method was developed for this purpose and permits a quality control
of polymer powders within a short time (Fig. 22).

The plastic powder to be tested is hydraulically compacted to a tablet under constant
conditions and pressed onto a spike on a test plate. The test plate so prepared is
placed into a stand at an angle of 45° and then heated in an oven at a predetermined
standard fusion temperature and time (for example 15 min.) The plastic powder tablet
melts and produces a so-called "snot-nose" with a certain characteristic length.
The shape and length of the "snot nose" is an important qualitative characteristic
of plastic powders.

The melting behavior and melt length determined by this test method are functions
of the particle-size distribution and of the viscosity of the plastic melt. Moreover,
this method can contribute to the optimization of the coating technology by ascertaining the optimal temperature for fusion.

6.   References

1.   Ward, J. L.;   Holidays in Thin Film Coatings: Detection, Cause,
     Logan R. J.:   Significance
                        Mat. Prot. & Perf. 11 (1972), 8, p. 19 - 22

2.   N. N.   Quality testing of the lining in oil-field pipes; internal
                        guidelines (German)
                        VETCO-Celle / 2.5.1978

3.   N. N.   N.A.M. requirements for evaluation and testing of internal
                        plastic coatings of tubing and drillpipes
                        (Revision B) / Jan. 1978

4.   N. N.: DIN 53 734: Testing of plastics / sieve analysis of powdered plastics
                        with air-jet sieve apparatus

5.   N. N.: DIN 53 203: Estimation of the fineness of dispersion by the grindometer

6.   N. N.: DIN 53 154: Falling-ball test on paint coatings and similar coatings

7.   N. N.: DIN 51 165: Determination of resistance to boiling water and water
                        vapour

8.   N. N.: DIN 50 903: Pores, occlusions, blisters and cracks

9.   Jorda, M. N.:   Paraffin Deposition and Prevention in Oil Wells
                        J. Petr. Techn. 18 (1966), p. 1605 - 1611

10. Schwenk, W.:   Parameters of creep corrosion of plastic coatings for
                        corrosion protection of tubular goods and their technical
                        significance (German)
                        gwf-gas/erdgas 118 (1977), p. 7 - 11

11. N. N.   NACE Standard TM-03-75: Test Method/Abrasion Resistance
                        Testing of Thin Film Baked Coatings and Linings using the
                        Falling Sand Method
                        October 1975

12. N. N.   ASTM D 968-15: Standard Method of Test for Abrasion
                        Resistance of Coatings of Paint, Varnish ,Lacquer, and
                        Related Products by the Falling Sand Method

13. N. N.: DIN 53 151: Cross-cut test on paint coatings and similar
                        coatings

14. N. N.   ASTM D 2794: Standard Test Method for Resistance of
                        Organic Coatings to the Effects of Rapid Deformation
                        (Impact) - Oct. 1969 -

15. N. N.: DIN 53 232: Measurement of the adhesion of paint coatings and
                        similar coatings according to the pull-off method
                        (April 1973)

16. Domalski, W.:   Corrosion protection of steel pipes with epoxy
                         powder coatings (German)
                        Shell-Pipe Coating Symposia / May 1978

17. N. N.   New duroplastic coating materials for extreme
                        thermal and chemical conditions (German)
                        Werkstoffe Korrosion 15 (1964) p. 995 - 1000

18. Voyles, J. L.:   Internal Tubular Plastic Coatings for Sour Gas and
                        Crude Service
                        SPE of AIME Symp. Sour Gas & Crude,
                        Tyler, Texas (11. - 12. 11. 1974)

19. Günther, E.:   Lining of steel pipes for the transport of gases
for public gas supply (German)
gwf-gas/erdgas $\underline{119}$ (1978), p. 456 - 463

20. N. N.   A Linearized Epoxy-Phenolic Formulation Recommended
for Higher Temperature and Extremely Corrosive
Environments (VETCO specification)

21. Hauk, V.;   Testing of coated pipes; Mannesmann research report
Meyer, W.;   646/1974 (German)
Schwenk, W.:

22. N. N.   Technical Practices Applicable to Internally Plastic
Coated Oil Field Tubular Goods for Corrosive Service
Report NACE TU-Committee T-1 G on Protective
Coatings and Non-metallic Materials for Oil Field
Use / Task-Group T-1 G-1 on Plastic Coatings for
Tubular Goods (Febr. 1966)

23. Moses, T. E.:   Baked-On Plastic Coatings
Proc. 8th Ann. West Texas Oil Lifting Short Course
(20.-21.4.1961), p. 105 - 108

24. N. N.   Plastics control paraffin buildup
Can. Petr. Eng. (June 1965), p 33 + 39

25. Ramey, W. J.:   Coated Pipe: A Practical Remedy for Paraffin Deposition
Drill Bit (August 1966), p 25 - 27

26. De Monye, G.:   Combatting paraffins in oil fields (German)
Erdöl Z. $\underline{74}$ (1958), p. 133 - 143

27. Thompson, L. J.:   Internal Coatings and Linings
(For Tubular Goods, Line Pipe, Tanks and Vessels)
Oklah. Univ. Corr. Contr. Short Course, Norman
(15.-17.9.1969), Proc. (1969), p.W 1 - W 29

28. Roberson, G. R.:   Comparison of Corrosion Rates: Wireline-Cut Plastic
Coated Oil Well Tubing vs. Wireline-Cut Bare Tubing
Mat. Perf. $\underline{13}$ (1974) 12, p.26 - 28

29. N. N.: DIN 50 125:   Tensile tests specimens: guidelines for preparation

30. N. N.: DIN 50 140:   Tensile testing of pipes and pipe strips without a
strain gauge (Sept. 1965)

Table 1: <u>Average tickness of epoxy-resin coatings on metallic test specimens</u>

Coating (Competitor)		Average coating thickness in μm
A-1	B	304
	R	262
A-2	B	314
	R	227
	R	206
B-1	B	340
	R	283
C-1	B	276
C-2	B	210
C-3	B	245
	R	200
C-4	R	274
C-5	B	126
	R	125
C-6	B	186
	R	122
E-1	B	315
	R	388
D-1	R	359

<u>Note:</u>

B: Flat sheet or plate specimen
R: Pipe segment

Table 2: <u>Curing tests in MEK and Acetone; pencil-hardness test</u>

Pencil hardness test

Coating (Competitor)	before testing	30 min in MEK	24 h in MEK	30 min in Acetone	24 h in Acetone
A-1	4 H	4 H	H	–	–
	4 H	2 H	–	–	–
	4 H	–	–	3 H	6 B
	4 H	4 H	–	–	–
A-2	2 H	H	–	–	–
	2 H	B	–	–	–
	2 H	–	–	6 B	–
B-1	H	5 B	–	–	–
	H	–	–	3 B	–
C-1	H	6 B	–	–	–
	H	–	–	6 B	–
C-2	2 H	6 B	–	–	–
	2 H	–	–	6 B	–
C-3	H	6 B	–	–	–
	H	6 B	–	6 B	–
C-4	H	6 B*	–	6 B*	–
	2 H	6 B*	–	6 B*	–
C-5	2 H	6 B	–	6 B	–
C-6	2/3 H	6 B	–	6 B	–
E-1	H	6 B	–	–	–
	H	–	–	6 B	–
	H	6 B	–	6/5 B	–
	H/2 H	6 B	–	6/5 B	–
D-1	2 H	H	–	H	–

<u>Note:</u>
* Very poor, pronounced swelling, coloration of solution

Table 3: <u>Bonding strength</u>

Coatings (Competitor)		Bonding strength in $N \cdot mm^{-2}$
A-1	B	: 600 – 650
	R	: 1100
A-2	B	: 600 – 750
	R	: 750 – 900
B-1	B	: 500 – 800
	R	: 900 –1000
E-1	B	: 400 – 600
	R	: 1000
C-1	B	: 750
C-2	B	: 450
C-3	B	: 500 – 550
	R	: 900 –1000
C-4	R	: 500
C-5	R	: 750
C-6	R	: 900
D-1	R	: 900

<u>Note:</u>

B: Flat sheet or plate specimen
R: Pipe segment

Table 4: <u>Abrasion test</u>

Coating (Competitor)		Loss of mass in g with 50 l sand	100 l Sand	Coefficient of abrasion in $mg \cdot l^{-1} \times 10^{-3}$
A-1	B	0,017	0,036	34, 36
	R	0,050	0,070	100, 70
A-2	R	0,040	0,070	80, 70
	R	0,050	0,060	100, 60
	B	0,024	0,048	48, 48
B-1	B	0,012	0,024	24, 24
	R	0,013	0,022	26, 22
	R	0,009	0,019	18, 19
	R	0,010	0,020	20, 20
	R	0,016	0,034	32, 34
E-1	B	0,037	0,047	74, 47
D-1	R	0,020	0,030	40, 30
C-1	B	0,018	0,038	39, 38
C-2	B	0,020	0,032	40, 32
C-3	B	0,016	0,033	32, 33
	R	0,020	0,080	40, 80
	R	0,030	0,080	60, 80
C-4	R	0,020	0,050	40, 50
	R	0,020	0,030	40, 30
C-5	R	0,030	0,110	60, 110
C-6	R	0,040	0,060	80, 60

<u>Note:</u>

B: Flat sheet or plate specimen
R: Pipe segment

Table 5: Direct-impact test (acc. to ASTM 2794-69) on coated pipe segments

Coating (Competitor)	10 N·m	18 N·m	20 N·m
A-1	a ++	a +	a +
A-2	c +	c -	c -
B-1	ok +++	a +	a +
D-1	ok ++	a -	a -
C-3	a ++	a ++	a ++
	ok +++	ok +++	a ++
C-4	b ---	b ---	b ---
	b ---	b ---	b ---
C-5	ok +++	a +++	a ++
C-6	ok ---	ok +++	a ++

Definition of symbols:

ok = no visible damage to the coating
a   = coating squeezed away
b   = damaged area extends beyond site of impact; coating badly damaged; insufficient elasticity
c   = coating chipped of down to bare metal
+++ = excellent coating; no significant damage
++ = slight damage to the coating
+   = pointwise damage down to base metal
-   = considerable damage down to base metal
-- = coating chipped off at immediate site of impact
--- = damage over large area extending beyond site of impact

Table 6: Breaking test on coated drill-pipes

Coating (Competitor)	Test results
A-2	Loss of adhesion only at the immediate site of fracture; however, alteration of coating at a considerable distance from site of fracture, otherwise ok
C-3	Loss of adhesion only at the immediate site of fracture, otherwise ok
C-4	Pronounced forming of transverse cracks over the entire specimen length; coating sensitive to tensile stresses; adhesion to the base metal very good; elasticity insufficient
C-5	Simultaneous fracture of plastic coating and steel; no loss of adhesion, only slight damage at the immediate site of fracture
C-6	Results as for competitor C-5
B-1	Complete exfoliation of coating over a large area extending far beyond the site of fracture; plastic coating can be further stretched but loses elasticity
A-1	Pronounced loss of adhesion at the site of fracture
D-1	Complete exfoliation of coating over a large area extending far beyond the site of fracture; inferior qualtiy of coating

Table 7: Torsional test on coated drill pipes

Coating (Competitor)	Evaluation of coatings
A-1	ok; no visible damage
A-2	ok; no visible damage
B-1	ok; no visible damage
D-1	ok; no visible damage
C-3	ok; no visible damage
C-4	Transverse cracks over large area
C-5	ok; no visible damage
C-6	ok; no visible damage

Table 8: Solutions employed for autoclave tests

Designation of solution	ph-value of original solution at 20 ± 2 °C	120 ± 5 °C	pH-value at the test temperature 135 ± 5 °C	150 ± 5 °C	desired pH-value
A1, B1, C1	6,1 ± 0,1	A1 = 7,0 ± 0,1	B 1 = 7,1 ± 0,1	C1 = 7,3 ± 0,1	7,0
A3, B3, C3	11,1 ± 0,1	A3 = 8,9 ± 0,1	B 3 = 8,6 ± 0,1	C3 = 8,2 ± 0,1	8,5
A2	12,2 ± 0,1	A2 = 9,8 ± 0,1			10,0
B2, C2	13,0 ± 0,1		B 2 =10,0 ± 0,1	C2 = 9,7 ± 0,1	10,0
A7, B7, C7	8,3 ± 0,1	A7 = 5,9 ± 0,1	B 7 = 5,6 ± 0,1	C7 = 5,2 ± 0,1	original drilling mud
A6, B6, C6	12,5 ± 0,1	A6 =10,3 ± 0,1	B 6 =10,0 ± 0,1	C6 = 9,7 ± 0,1	10,0
A5, B5, C5	11,5 ± 0,1	A5 = 8,8 ± 0,1	B 5 = 8,5 ± 0,1	C5 = 8,2 ± 0,1	8,5
A4, B4, C4	10,2 ± 0,1	A4 = 7,4 ± 0,1	B 4 = 7,0 ± 0,1	C4 = 6,6 ± 0,1	7,0

Table 9: Cross-cut test following the autoclave test, acc. to DIN 53 151
(3 % NaCl solution / pH-adjustment with HCl or NaOH)

Coating (Competitor)	Solution at pH 7,0			Solution at pH 8,5			Solution at pH 10,0		
	A 1 120 °C	B 1 135 °C	C 1 150 °C	A 3 120 °C	B 3 135 °C	C 3 150 °C	A 2 120 °C	B 2 135 °C	C 2 150 °C
A-1	+	+	+	+	+	+	+	+	+
A-2	+	+	+	+	+	+	-	--	---
B-1	+	+	+	+	+	+	+	+	+
C-1	+	+	+	+	+	+	+	+	-
C-2	+	+	+	+	+	+	+	+	+
C-3	+	+	+	+	+	+	+	+	+/-
C-4	+	+	+	+	+	+	+	+	+
C-5	+	+	+	+	+	+	+	+	+
C-6	+	+	+	-	+	+	--	+	+
D-1	+	+	+	+	+	+	-	-	-
E-1	+/-	+/-	+/-	+	+/-	+/-	--	--	-

Definition of symbols:

+ = Gt 0 C; no objections

- = Gt 0 C - Gt 1 C; slight objections

-- = Gt 1 C - Gt 2 C; coating brittle, insufficient adhesion

--- = Gt 3 C; coating very brittle with almost total loss of adhesion

Table 10: Cross-cut test following the autoclave test, acc. to DIN 53 151

(alkaline-earth drilling mud / pH-adjustment with HCl or NaOH)

Coating (Competitor)	Original drilling mud			Drilling mud pH 7,0			Drilling mud pH 8,5			Drilling mud pH 10,0		
	A 7 120 °C	B 7 135 °C	C 7 150 °C	A 4 120 °C	B 4 135 °C	C 4 150 °C	A 5 120 °C	B 5 135 °C	C 5 150 °C	A 6 120 °C	B 6 135 °C	C 6 150 °C
A-1	+	+	+	+	+	+	+	+	+	+	+	+
A-2	+	+	+	+	+	+	+	+	+	+	+	--
B-1	+	+	+	+	+	+	+	+	+	+	+	+
C-1	+	+	+	+	+	+	+	+	+	+	+	+
C-2	+	+	+	+	+	+	+	+	+	+	+	+
C-3	+	+	+	+	+	+	+	+	+	+	+	+
C-4	+	+	+	+	+	+	+	+	+	+	+	+/-
C-5	+	+	+	+	+	+	+	+	+	+	+	+
C-6	+	+	+	+	-	+	+	-	+	-	--	--
D-1	+	+	+	+	+	+	+	+	+	+	+	---
E-1	+	+	+	+	-	+	+	+	+	+	+	+

Definition of symbols:

+ = Gt 0 C no objections
- = Gt 0 C - Gt 1 C slight objections
-- = Gt 1 C - Gt 2 C coating brittle, insufficient adhesion
--- = Gt 3 C coating very brittle with almost total loss of adhesion

Table 11: <u>Acid-resistance test (acc. to DIN 51 167)</u>

Pencil hardness test

Coating (Competitor)	before testing	80 °C/liquid	80 °C/gas phase	100 °C/liquid	100 °C/gas phase
A-1	4 H	2 H	2 H	2 H	2 H
A-2	4 H	2 H	2 H	2 H	3 H
B-1	6 H	3 H	3 H	H	2 H
E-1	2 H	H	H	H	H
D-1	4 H	4 H	4 H	4 H	4 H
C-1	H	H	H	H	-*
C-2	2 H	2 H	2 H	2 H	2 H
C-3	4 H	2 H	2 H	H	H
C-4	3 H	H	H	H	-*
C-5	4 H	3 H	3 H	3 H	2 H
C-6	4 H	3 H	3 H	3 H	2 H

<u>Explanation:</u>

* Coating exfoliated; no measurement of hardness possible

Table 12: <u>Water-vapour test (acc. to DIN 51 165)</u>

Pencil-hardness test

Coating (Competitor)	before testing	after testing
A-1	4 H	4 H - 2 H
A-2	5 H - 4 H	5 H - 2 H
B-1	6 H - 2 H	4 H - H
D-1	4 H	4 H
C-1	H	H
C-2	2 H	2 H
C-3	4 H - 2 H	2 H - H
C-4	3 H	2 H
C-5	4 H	3 H
C-6	4 H	3 H

Table 13: <u>Surface roughness</u>

Coating (Competitor)	Surface quality
A-1	+++
A-2	+/-
B-1	+/---
C-3	++
C-4	++/-
C-5	++
C-6	+++
D-1	--

<u>Definition of symbols:</u>

+++   very good surface

++   good surface quality

+   satisfactory surface quality

-     barely sufficient surface quality

--   unsatisfactory surface quality

---   inferior surface quality

For suppressing paraffin deposition on the pipe wall only
the qualities with +++ or ++ are suitable

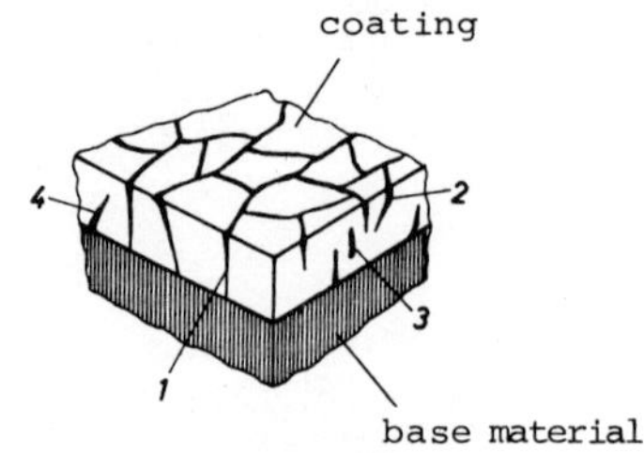

1 crack passing through coating
2 open discountinuous crack (surface crack)
3 closed (hidden) crack, surrounded by coating material
4 closed (hidden) crack, bounded by base material

Fig. 1: Cracks in a plastic coating

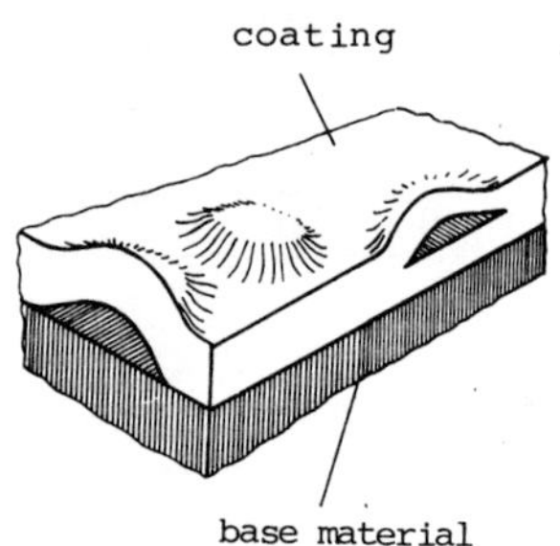

Fig. 2: Blisters in a plastic coating

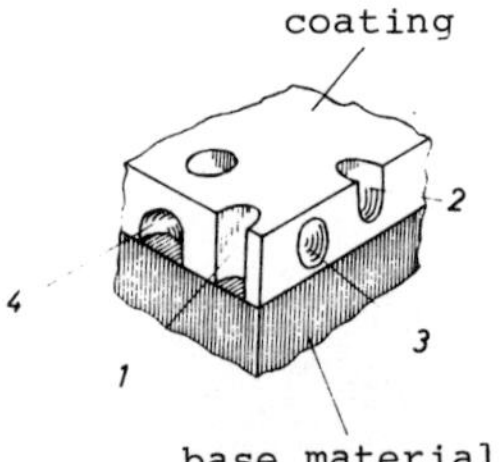

1 holiday (pore passing through coating)
2 pit (open discontinuous pore)
3 closed (hidden) pore
4 closed (hidden) pore, bounded by base material

Fig. 3: Holidays (pores) in a plastic coating

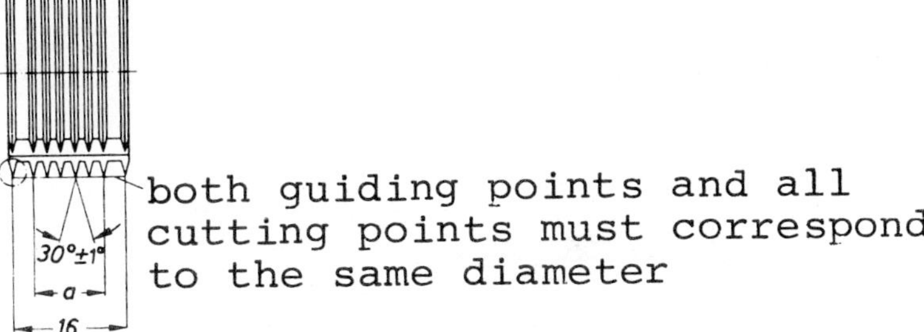

both guiding points and all
cutting points must correspond
to the same diameter

dimensions in mm

<u>Fig. 4</u>: Cutters of a multi-cutter device for the cross-
       cut test

cross-cut-value	description	illustration
Gt0	The edges of the cross-cut are completely smooth; no part of the coating has chipped off	—
Gt1	Small chips of the coating have chipped off at the intersections of the cross-cut lines;chipped area approx. 5 % of the tested area	
Gt2	The coating has chipped off along the section edges and/or at the intersections of the cross-cut lines; chipped area approx. 15 % of the tested area	
Gt3	The coating has chipped off along the section edges, partially or totally in wide strips and/or has totally or partially chipped off from individual sections; chipped area approx. 35 % of the tested area	
Gt4	The coating has chipped off in wide strips along the section edges and/or totally or partially chipped off from individual sections; chipped area approx. 65 % or more of the tested area	

<u>Fig. 5</u>: Evaluation table for the cross-cut test

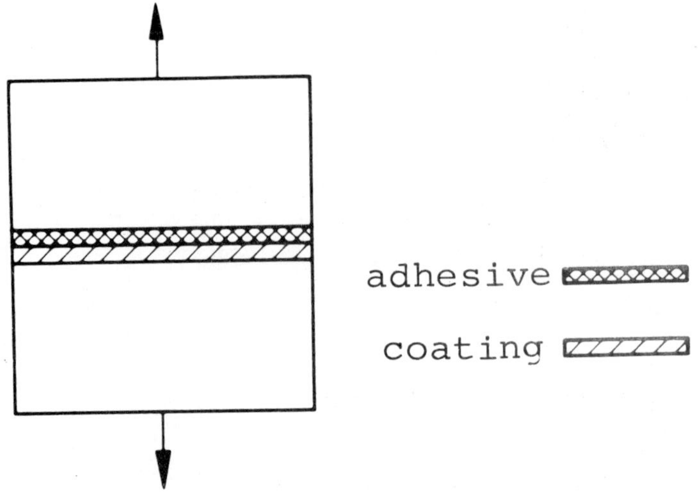

Fig. 6: Pull-off test (opposite   cemented test plugs)

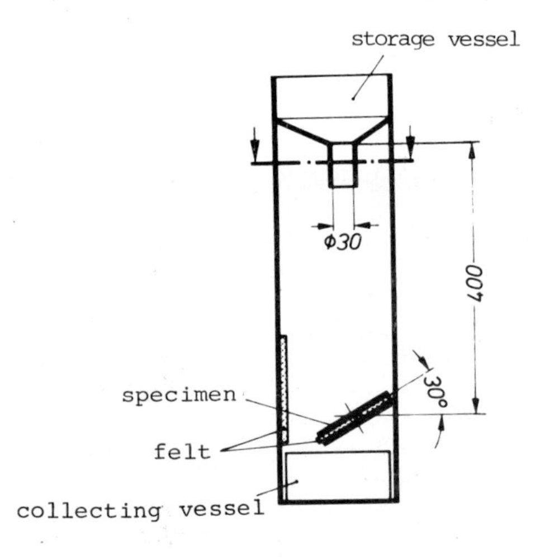

Fig. 7: Abrasion test with falling sand

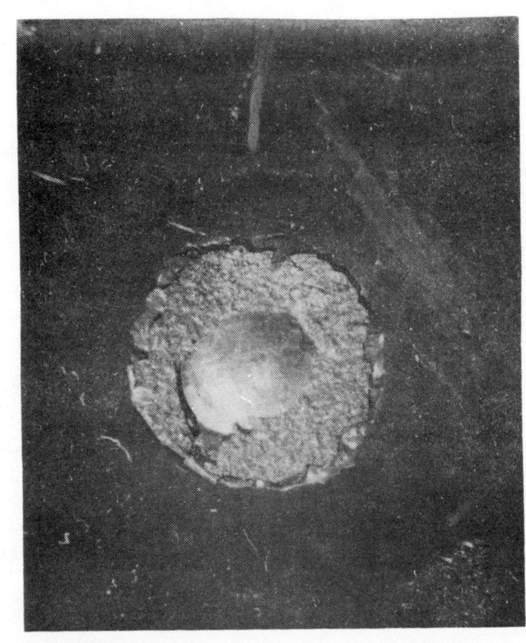

Fig. 8: Direct impact test with
chipping over extensive area

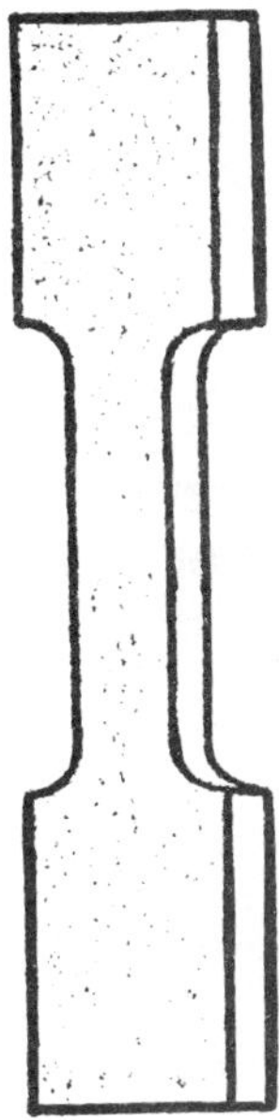

Fig. 9: Specimen for tensile strength and breaking test

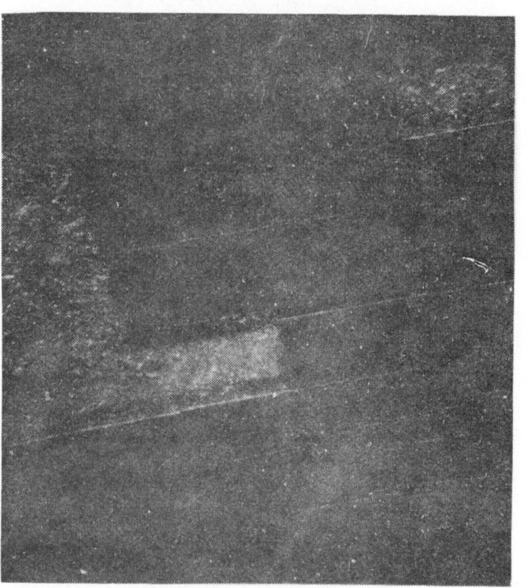

Fig. 10: Tensile test within the elastic range; brittle coating

Fig. 11: Breaking test; good coating

Fig. 12: Breaking test; inferior coating

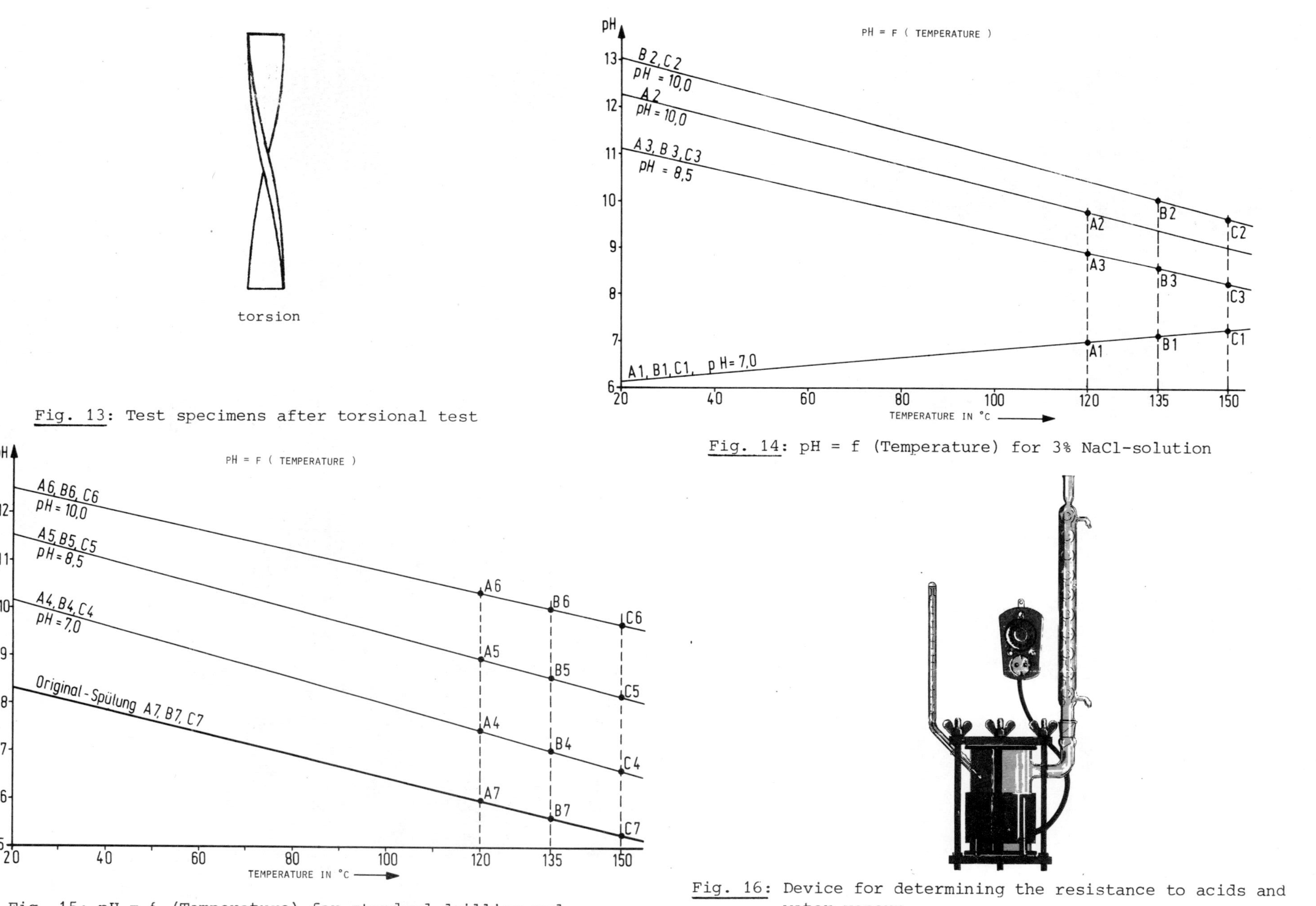

Fig. 13: Test specimens after torsional test

Fig. 14: pH = f (Temperature) for 3% NaCl-solution

Fig. 15: pH = f (Temperature) for standard drilling mud

Fig. 16: Device for determining the resistance to acids and water vapour

44

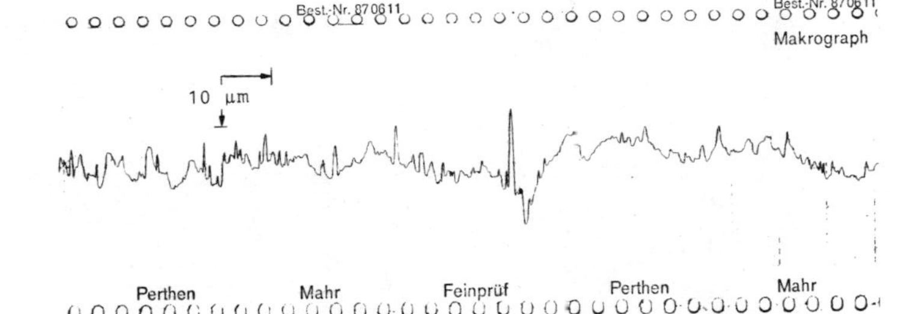

Fig. 17: Roughness profile; inferior coating

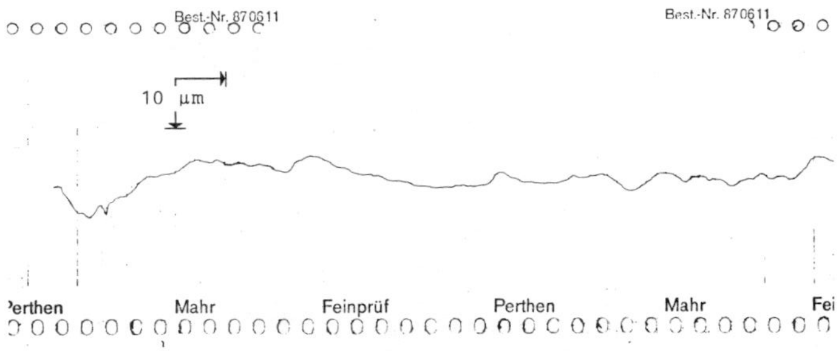

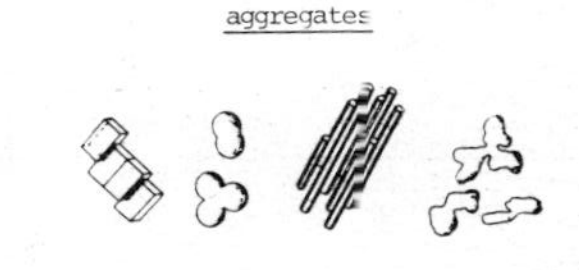

Fig. 19: Possible grain shapes for the sieve analysis of
polymer powders

Fig. 18: Roughness profile; good coating

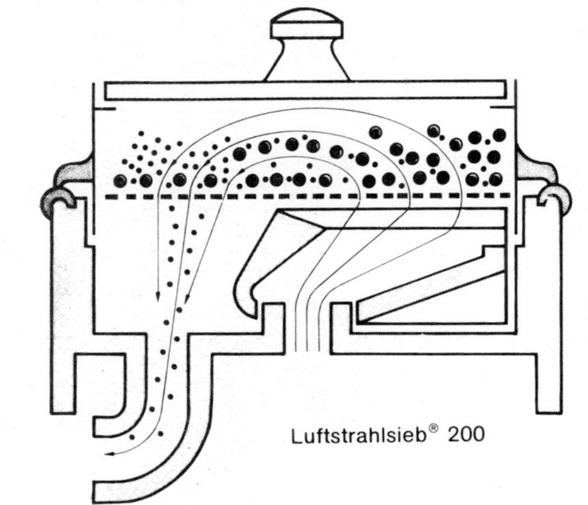

Fig. 20: Schematic diagram of the ALPINE[R] air-jet sieve 200

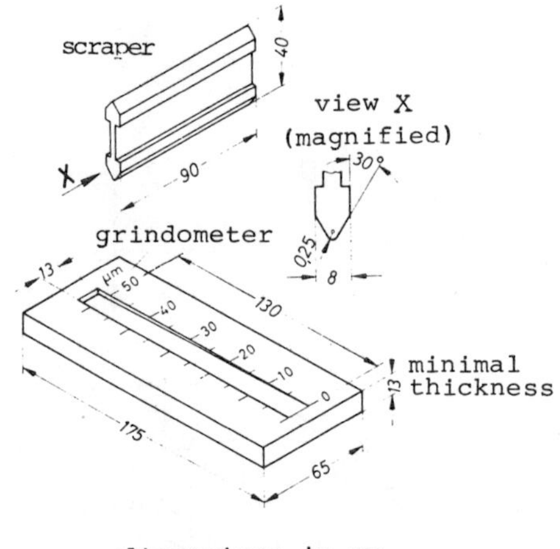

Fig. 21: Grindometer test acc. to HEGMAN

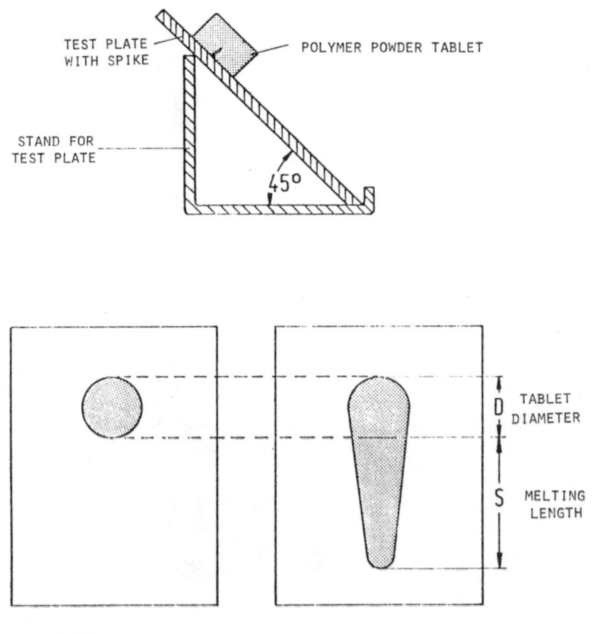

Fig. 22: Test of polymer powders by the fusion method

internal and external
protection of pipes

September 5th - 7th, 1979

PIPELINE COATING EVALUATION TEST METHODS
AND THEIR RELEVANCE TO PERFORMANCE

E.E. Hankins

Long Products, U.K.

Summary

The importance of measuring the potential performance of a pipeline coating is obvious as the economics of recovering a corroded pipeline are very unattractive and, in some cases, potentially dangerous. There are many and various test methods available and employed by authorities and manufacturers. This review compares these methods to identify their value in evaluating a pipeline coating.

Shrier[1] considered the following requirements as the required properties of a pipeline coating:-

1. Ease of application.
2. Good adhesion to pipe.
3. Good resistance to impact.
4. Flexibility.
5. Resistance to soil stress.
6. Resistance to flow.
7. Water resistance.
8. Electrical resistance.
9. Chemical and physical stability.
10. Resistance to soil bacteria.
11. Resistance to marine organisms.

These appear to be comprehensive with one notable omission viz. resistance to cathodic disbondment which would add twelve basic requirements. How the basic needs can be defined in greater detail and how they can be measured has to be considered.

Held at Imperial College, London, England.
Organised and Sponsored by BHRA Fluid Engineering

1.   Ease of application: Obviously the different coating forms available need to
be measured by different methods as far as their application and efficiency is
concerned.   With a hot applied enamel its rheology at coating temperature would be
important, whereas with extruded polyethylene coating this would be evaluated by its
melt-flow-index.   Alongside these considerations the practicalities would have to
be taken into account.   For example, it would be extremely difficult to extrude
polyethylene on a pipe over the ditch and the need to provide heating for the enamel
is a restriction for field application.   With cold applied tapes these practical
problems do not arise but, equally, here are no simple tests for measuring the
application properties such as viscosity or MFI.

What properties are expected of the tape during the wrapping process? Because
the tape is to be wrapped spirally around the pipe and to overlap itself these
combine to cause one side the tape to cover a greater circumference.   Thus the tape
has to be stretched more on one side than the other and this is achieved applying
more tension on one side.

The ability of the tape to do this is determined by its elongation, its modulus
and yieldpoint.   The latter is important as the force required to distort the tape
should not be such that the yield point is exceeded.   Most specifications ignore
this and account for tensile strength and elongation at break which, whilst important,
in other aspects do not necessarily reflect conformability.   A minimum yield point
of 11 kg per 25 mm width is felt necessary with ratios of yield to ultimate strength
of between 1.5 and 2.   If this ratio is exceeded full utilisation of the tape
strength cannot be made.   Yield point is dependent upon two factors.   The type of
plastic and its thickness.   PVC will exhibit a higher yield point than PE due to
its elastomeric nature.

2.   Adhesion to pipe: It is very clear that the adhesion to the pipe is an
essential prerequisite to corrosion protection but care must be taken in assessing
adhesion and in specifying the minimum requirement.   Adhesion can be measured by
shear, peel or tensile force and the industry generally relies on peel.   Possibly
this is because peel is easily achieved in the field although, due to the complexity
of the forces involved, frequently leads to variable results when tested under
laboratory conditions.   There are two main methods of measuring peel viz. 180° peel
and a 90° peel and most national standards apply both as exampled by DIN 30672.
The principles involved are the same for all methods.   The 180° consists simply of
applying a tape to flat metal plate and peeling that through angle of 180° at a
fixed speed of 300 mm. per minute.   The 90° peel employs a rotating drum in which
the tape is pulled away from the drum at 300 mm. per minute, the drum rotating to
maintain an angle of 90°.   The same force is not required for both tests due to the
complexity of forces involved in peel adhesion.   The main difficulty with peel is
that it is not the only force that the bonded coating is required to resist and
equally important is shear which is the main force imposed by soil stress.
Measurements of shear strength reveal interesting results, some products showing
high peel strengths when exposed to shear forces do not perform well.   There is no
standard method for measuring shear adhesion of coating but by fixing rigid plates
to the coating good results can be achieved.   Interpretation of the degree of bond
required is somewhat varied and care should be taken in the thinking that high bond
strengths is all that is required.   The mode of failure is also important and it is
advantageous that the adhesive should break cohesively, rather than an interfacial
failure exposing the pipe, to allow some coating to remain in contact with the pipe
thus providing protection.

3.   Good resistance to impact: Damage to coatings is reported to be one of the
major causes of corrosion[2] and therefore resistance to impact is of major import.
The normal method of measurement is to drop a known weight through a pre-determined
distance and observe the effect.

ASTM G13 limestone drop test consists of dropping, through a chute onto the
coated pipe, 16 kilograms of stones and measuring any holiday caused.   The stones
are dropped through height of 2 metres and impacting is continued till failure
occurs.   The number of drops is noted and recorded.

ASTM G14 consists of dropping a 15 mm dia. tup weighing 1.36 kg. through
distances up to 1.22 metres, continuing to increase the impact energy by increasing
the height until rupture of the coating takes place.   If necessary, a low voltage
wet sponge holiday detector can be used as identifying breaks.   DIN have approached
the assessments somewhat differently by specifying 3 classes of impact force, namely
2 Nm, 8 Nm and 15 Nm using a dropping height of 1 metre and 25 mm tup but the
assessment is based on the need for the impacted area to maintain a lielectric
strength of 5 kV per mm plus 5 V.

It is clear when comparing the results of the two methods an account should be
made of the differences as the DIN method the coating is still capable of providing
a total corrosion protection whereas with ASTM no such guarantee is present.   The
ASTM is therefore of greater value in assessing differences between coatings as a
research tool but from the users point of view DIN ensures that following impact the
coating is intact.   A practical advantage with DIN is the use of a fixed height
ensuring greater accuracy for variations in height are more difficult to control than
changes in weight.

Some methods introduce the use of a chisel tup rather than a sphere but
considerable difficulty has resulted in the use of chisels because of radii
variations and direction of impact.   Some works have reported that differences
result using exactly the same coatings dependent upon whether the chisel drops
parallel to the pipe direction or at right angles to it, and they have also reported
that some coatings give superior results with chisel impact compared to spheres
whereas the reverse was true on other coatings.   This leaves the thought that
coatings might need to be selected as to whether the backfill or source of impact
was to be sharp or blunt provided it is not some hidden function of the test causing
differences.

One other area which has not received much attention is the effect of
temperature on impact and recent work undertaken shows little change in impact
resistance occurs at low temperatures but serious changes occur above 50°C and
further work is being undertaken.   Another investigation is the effect of plastic
outerwraps on impact.   Recent examination of some outerwraps have led to the
disturbing results that although the outerwrap is unaffected the primary wrap has
been fractured.   This leads to false security as neither visual nor electrical
examination would bring the fault to light and does open up the need to examine the
outerwrap concept in some detail as full protection may not be achieved with current
products.

4.   Flexibility: Shrier defined it as the ability of the coating to withstand
bending deformation and to absorb the effects of thermal changes.   ASTM G10 covers
bendability of coated pipe using small dia. pipes if approximately 25 mm and bending
in several stages through a frame via variable radius mandrel then examining the
bent section for fractures by wet sponge holiday detector.

Flexibility is a function of extensibility and adhesion, and whilst failures
have occurred in some enamel coatings and in the development stages of epoxy powders
and polyethylene extruded coatings, cold applied tapes exhibit sufficient
extensibility and adhesion to withstand bending moment.

5.   Resistance to soil stress: As we mentioned earlier, some of the effects of
soil stress can be measured by the shear adhesion but surprisingly there are not any
soil stress tests in the national or international test methods.   British Gas have
produced a test method which consists of immersing a pipe in a standard clay bed and
wetting the clay, then pumping hot water at 50°C through the pipe causing the clay
to dry under controlled conditions.   On drying the clay will shrink thus imposing
a stress on the pipe coating.   The cycle is repeated 10 times examining the pipe at
each cycle completion.   British Gas have scaled up these tests to use 36" dia.
pipes and correlated the results successfully.

ASTM D427 measures the Soil Shrinkage factor and use of this would help in the
coating selection by avoiding the use of those coatings subject to stressing in
areas of high shrinkage ratios.

Yield point of the coating and the soil adhesion will influence the coating resistance, but in making the coating more resistant to the shear form some sacrifice to flexibility must be made. This balance needs to be carefully controlled.

Due to polyethelene having a lower yield point than PVC there is an increased danger that the movement takes place in the coating for a given stress resulting in greater wrinkling. Although this will be influenced upon the shear strength of the adhesive.

Problems with soil stressing can sometimes take an unusual turn, for example a vertical pipe buried in an old tidal basin was found after three years to have had the coating rolled down its length. Examination showed that high soil stresses were imposed in certain soil conditions causing significant downward thrust. This is a real problem when piling meets such soil conditions and the piles can be drawn further down into the soil and work by Shell[4] developed a grade of bitumen to act as a slip layer to negate the downward thrust. The coating can serve the same function with less happy results and this must be accounted for with vertical pipework providing high shear resistance.

6.   Resistance to flow: Shrier obviously was thinking of hot enamel coatings but all thermoplastic coating and some thermosets in fact will flow at sufficiently high enough pressures and/or temperatures. Measuring is not easy however and as with most pipes we are considering a buried situation when soil pressures are present and an impression or penetration test is a good way of assessing this flow.

DIN 30672 is a good test method of assessing this by applying steady loads to a coated pipe for 72 hours and measuring the impression depth. Three classes of weight are involved and DIN requires that not greater than 25% of the original thickness will be impressed or that at least 0.6 mm thickness should remain. The test can be dealt with at varying temperatures quite simply although DIN only require the test to be carried out at 23° and in certain instances $50^{o}C$.

ASTM G17 also cover penetration but the published method does not provide details of the weight loading or surface areas of the rods. The three classes established by DIN are felt to be widely spread and the inclusion of two additional classes would be helpful. Some newer specifications such as the Dutch National Standards employ the DIN test but use only one loading which provides for less selection.

7.   Water resistance: As water is one of the main constituents to keep away from the pipeface the ability of the coating to resist water is important. There are two properties somewhat, though not entirely, inter-dependent that affect the efficiency of the coatings and their anti-corrosive properties, water absorption and water vapour transmission.

As regards the former, most national and international standards call for a weighed specimen to be immersed in distilled water for a set period removed, dried and reweighed. Differences between the tests are ASTM D570 has an immersion time varying from 2 hrs to 2 weeks. BS 2782 fixes on 24 hrs but expresses its results not in percentage increase but in milligram increase. They all take a simple specimen and immerse it for a specified period. Some results fail to express the immersion time and if 2 hrs. is used it will provide significantly lower results than a 24 hr. immersion. So here care in interpretation of results must be exercised and the necessary questions asked before comparing results. British Gas have modified the test methods by taking pieces of the specimens and then joining, in the case of tapes, the two adhesive faces together and immersing this fixed tape in tap water at $50^{o}C$ until a constant value is achieved during 24 hourly weighings.

ASTM G9 measures the penetration of water into a coating by changes in dielectric strength expressed against time it forms a measure of deterioration and provides a possible determination of coating thickness requirements. Water vapour transmission can be measured by a wide number of methods, all maintaining the basic system, the differences being to measure the property in differing situations.

Basically, a film of the material is placed to cover a small tray.   This is then
sealed around the edges to prevent the passage of moisture or vapour other than
through the coating.   The tray is filled with either a dessicant or water.   In the
latter case the assembly can be upright or inverse so as to bring the water into
contact with the coating.   The assemblies are then placed in an atmosphere of
controlled temperature and humidity and the weight loss in the case of water filled
trays or the weight gain in the case of dessicant filled trays is measured.   The
temperature and humidity levels are usually set at 25°C and 75% relative humidity
and 32°C and 90% relative humidity to cover the temperate and tropical conditions
respectively.   Measurements are taken at 24 hourly intervals until a steady rate
of weight change is achieved for six consecutive readings.   Calculation is then
made to express results in terms of grams per metre squared per 24 hours.

Obviously the lower the transmission rate the better but all coatings will
allow some passage of water vapour to take place and therefore the level at which
a dangerous corrosive situation can be reached must be set.   To do this it is
necessary to decide which conditions are to be used from the differing methods
available.

The main test method employed is ASTM E96 which has six variants of
application and method.   We consider that method A, BW & D are the relative
procedures for pipeline coatings but frequently find that in literature, more
alarmingly in specifications, a result or a constraint is given but the test method
is just quoted as ASTM E96 with no reference as to the method.   Water vapour
permeability rates of greater than 2 grams per m^2 per 24 hours would be considered
to be a maximum level.

Corrosion is dependent not only upon the presence of water but also of oxygen.
In fact, because the amount of oxygen present compared to water vapour is low the
rate of corrosion depends principally on the rate of diffusion of oxygen.   Oxygen
diffusion rates of plastic films of 1 mm thick suggest that corrosion rates of
1 micron per year provide an estimated life of 50 years[5].

One area that does not appear to be taken into account is the effect of ageing
on coatings in relation to gas permeability.   ASTM E154 goes some way towards this
by measuring water vapour transmissions after a wet and dry cycling plus long term
soaking.   In view of the finding on gas pipeline explosions in the United States [6]
there is a need to investigate the effects of thermal ageing on water vapour
transmission properties but have no evidence of this work being done.

8.   Electrical Resistance: This is generally a straightforward procedure of
measuring electrical resistance of the coatings.   Some procedures vary, for example
BS 2782 requires that a 24 hour pre-conditioning in distilled water before testing
whilst others such as ASTM D1000 only requires pre-conditioning at test conditions.
The methods usually measure surface resistivity and volume resistivity.   We favour
the latter being more meaningful.   It is usually considered that a volume
resistance in the order of 10^{10} M Ohms are required to insulate steel against the
effects of electrolytic corrosion.   DIN 30672 employs a variation by measuring
specific resistance.   This does appear to produce more consistent results.

Whilst not strictly electrical resistance the dielectric strength can well be
considered under this heading.   Again test procedures are relatively
straightforward according to BS 2782 or ASTM D1000.   Strengths are usually expressed
in kV/mm and minimum figures of around 20 kV/mm are required.   Obviously thicker
coatings will have higher breakdown strength in practice although they may only have
the same strength per unit thickness.

9.   Chemical & Physical Stability: Shrier identified this as the ability of
the coating to absorb low molecular weight substances or hardening due to oxidation.
There are therefore two areas to consider, the resistance to chemicals likely to
come into contact with the coating and the effects of thermal ageing.

ASTM G20 covers the test method for chemical resistance and consists of
immersing a coated pipe in a jar containing the reagent.   The coated pipes are

capped to prevent penetration at the ends and a 6mm holiday is made in the coating.
The coatings are examined at 6, 12 and 18 monthly intervals for visual signs for
decomposition, softening, swelling, blistering, cracking or solubility.   The
holiday area is examined for loss of adhesion.   It is also possible to examine the
coating for change in mechanical properties which does enable the results to be
quantified.

DIN 30672 covers the effects of thermal ageing applicable to tapes only.
The test is conducted at 100°C for 100 days and the change in tensile strength and
elongation is measured.   Changes of not greater than 25% are allowed.   British Gas
has a similar test carried out at 50°C for 200 days and measuring changes in tensile
strength, elongation, weight loss and adhesion.   As indicated earlier changes in
water vapour transmission should be included.

<u>10.   Resistance to Soil Bacteria</u>:   Here again national standards are lacking.
DIN 30672 requires assurance that coatings having saponification values in excess of
10 mg/KOH are resistant to bacteria but do not specify the method.   GAZ de France
and British Gas both have methods.   BGAM 9.26.13 consists of immersing specimens
of the coating in liquid media containing selected micro-organisms at 30°C for one
month.   The degree to which the coating will support bacteria is then measured.

Coal tar enamels are usually considered inherently resistant to micro-organism
attack[7] but other coatings are formulated to prevent attack by the inclusion of
bacterio-stats.   Coatings have been known to be attacked by micro-organisms and
bacterial corrosion due to sulphide producing bacteria is well known.   It is
therefore important to ensure that coatings are resistant to micro-biological attack.

<u>11.   Resistance to marine organisms</u>:   Shrier referred to the need to resist
molluscees and similar marine systems but as most underwater pipe coatings are clad
in concrete to provide anti-buoyancy this does not seem to be a particularly vital
requirement.   Experience in the North Sea has shown that the hazard from anchors
and fishing trawl boards is such that marine organisms pale in comparison.

<u>12.   Resistance to cathodic disbondment</u>:   There appears to be, at least by
omission, a divergence of import regards the hazards of cathodic disbondment.   That
it occurs is not in dispute but testing for its effect is sometimes fraught with
problems and comments[8] that ASTM G8 is not reliable have been made.   In view of the
general acceptance that there is a hazard we can only presume that the reason why so
few specifying authorities include any requirement for cathodic disbondment is due
to the difficulty of assessment.

ASTM seems to be the only national standard that have produced a test for
neither British Standards nor the relatively young DIN 30672 include such procedures.
British Gas have two tests, one of which is based on ASTM G8 Method A with
minor changes in the constitution of the electrolyte solution and another which
monitors the current demand and potential as with Method B, but in addition also
tests a specimen free from holidays using a potential of -3 V ($Cu/CuSo_4$).   ASTM
considers a repeatability and a reporoducibility of 25 mm is acceptable but there
is evidence that these are not necessarily easily achieved.

It is known that surface preparation plays a prominent part and after
comparing differing preparation methods have concluded that shot blasting to SA $2\frac{1}{2}$
is the only satisfactory and reliable procedure.   Recent work[9] has shown irregular
results, electrical monitoring showing no relationship between current demand and
area of disbondment, there is some evidence that this may be linked with the geometry
of the electrolyte solution vessel.   In view of the monitoring procedure of the
pipeline by current demand further work is needed to explain these effects.

One problem with cathodic disbondment testing is the long testing time.   This
becomes a problem in quality control testing and procedures have been developed to
provide short term testing of 6 days using impressed potentials of 4 volts.   Good
correlation with ASTM G8 has been achieved.

There are other test methods involving cathodic disbondment.   ASTM G19 covers

52

the disbonding characteristics of pipeline coatings in the buried state.   In this
case short length pipes are coated and predamaged with a 9.5 mm dia. holiday, buried
leaving 300 mm above the ground in a prepared site to take pipes and anode.   The
holes are refilled with soil or water soil slurry.   Soil resistance and the initial
pipe to soil potential is measured, and the anode is connected to one specimen and
the voltage and resistance measured at 1 monthly intervals for 18 months and the
area of disbondment then measured.   This method may have value for pre-determining
the efficacy of different coatings in one soil system though the results in one soil
would not necessarily be relevant to another.

ASTM G42 approaches the problem from the angle of elevated and cycled
temperatures.   The test is in two parts; (a) a constant elevated temperature of
60°C using ASTM G8 as the basic procedure whereas (b) runs as for (a) for between
4 and 7 days and then allows the specimen to cool while maintaining the electrical
stress.   After $3\frac{1}{2}$ hours the specimen is removed, washed and placed in a cold
chamber at 15°C for 2 days.   The cycling is then repeated until a period of 30
energised days are completed and the disbonded area is measured.   Although these
tests exist they do not appear to be greatly used in practice so their correlation
is not known

The causes of cathodic disbondment are thought to be dependent upon four
inter-related phenomena.

They are (1) electro-osmosis (2) coating holiday sizes (3) A.C. discharge
(4) underfilm moisture.

Of these electro-osmosis is thought to play a part equal to that of hydrogen
liberation in causing coating disbondment.   It is based on the fact that a
potential applied to a porous diaphraghm will cause water to flow through the
capilleries of the diaphraghm and the force that pushes water away from the anode
is similar to that forcing water towards the cathode.   Therefore water can be forced
through a semi-permeable membrane resulting in disbondment.   The coating holiday
sizes is more straightforward as for a given voltage and resistivity current density
will vary inversley with the radius of the holiday so a decrease in holiday size will
therefore increase current density.   The A.C. voltage and trapped moisture do not
appear on the evidence available to play a major part in disbondment although have
inter-relationship with the other two features.

<u>Joints & Repairs</u>: Protection of joints and fittings in coated pipelines is
considered in ASTM G18 by measuring leakage current, capacitance and dissipation
factor to indicate changes in coating effectiveness.   Additionally note must be
taken of the adhesion between the factory coatings and the on-site applied coatings.

<u>Stress Corrosion Cracking</u>:   The problems met with stress corrosion cracking
in the U.S.A. and Iran have led to test procedures being established by Battelle
and the University of Newcastle   although it is still too early for a firm
standard to be produced.

The procedure consists of applying a fixed strain rate to coated pipe line
steel specimen including a holiday.   The specimens are immersed in an environment
in sodium carbonate and sodium bicarbonate solutions and the crack velocity is
measured.   The resultant fracture is examined for intergranular cracking.   This
test procedure has been found very helpful in the evaluation of the performance of
anti-stress corrosion cracking primers.

Having reviewed the areas of coatings that appear important and as always
with such reviews we see weaknesses in both test procedures and interpretation.
Most test procedures suffer from some weaknesses in attempting to simulate
conditions on an accelerated time scale whilst introducing reduced scale in the
operation.   Careful judgement of the test method is therefore needed together with
monitoring performance in the field which in itself will improve correlation.
This in turn will lead to improvements in testing procedures and such as we have
reviewed in this talk will eventually lead to higher performance products.

References

1. Shrier. Corrosion. Vol.2. 2nd Edit. 1976 p.15.82.

2. H.S. Edwards. B.H.R.A. Conference Canterbury 1977 Proceedings.

3. H.S. Edwards. Quality Assurance Aspects of Coatings BHRA 1977.

4. Shell Bitumen Review No.35 July 1971.

5. Proteccion Anticorrosiva de Tuberias. Peter Baseler.

6. R.R. Fessler. Pipeline Industry March 1976 page 37.

7. E. Pankhurst. BHRA Conference Canterbury 1977.

8. G.M. Harris. Recent Developments and the Performance of Tape Coatings.

9. Private Communications.

10. R.N. Parkins.
    R.R. Fessler.   Mat. in Engineering Applications Vol.1. Dec. 1978.

11. R.N. Parkins & R.D. Tems. NACE Corrosion 1979.

Fig. 1   Erosion rates of metallic iron surfaces

SITUATION	mm/year	g/m^2/year
1) Open country	0.001 – 0.01	7 – 70
2) Open industrial	0.01 – 0.05	70 – 350
3) Soil natural	0.01 – 0.05	70 – 350
4) Soil corrosive	0.05 – 0.2	350 – 1400
5) Soil anaerobic	0.1 – 0.4	700 – 2800
6) Cold fresh water	0.02 – 0.05	140 – 350
7) Hot fresh water	0.05 – 0.1	350 – 700
8) Calm salt water	0.05 – 0.2	700 – 1400
9) Rough salt water	0.2 – 1400	1400 – 3000

Fig. 2   Effect of temperature on impact resistance

PIPE TEMPERATURE:	IMPACT FORCE:	DIELECTRIC STRENGTH:		DIELECTRIC REQUIREMENT:	COMMENT:
70°C	8Nm	9 Fail 6 kv	1 Fail 8 kv	16 kV	Fail Class B.
50°C	8Nm	6 Fail 8 kv	4 Fail 10 kv	16 kV	Fail Class B.
30°C	8Nm	10 Pass 16 kV		16 kV	Pass Class B.
10°C	8Nm	10 Pass 16 kV		16 kV	Pass Class B.
-10°C	8Nm	10 Pass 16 kV		16 kV	Pass Class B.
70°C	2Nm	10 Pass 16 kV		16 kV	Pass Class A.
30°C	15Nm	10 Pass 16 kV		16 kV	Pass Class C.
-10°C	15Nm	6 Pass 10 kV 4 Pass 8 kV		16 kV	Fail Class C.

Fig. 3   Variants of ASTM D.96.

METHOD:	SUGGESTED SITUATIONS:	TEST CONDITIONS:
A	Materials for use at low range of humidities.	Desiccant at 23° C. 50% RH.
B	Materials at high range of humidities but not wetted.	Water at 23° C. 50% RH.
BW	Materials wetted on one surface but when hydraulic head is unimportant.	Water inverted at 23° C. 50% RH.
C	High Temperature with low range humidities.	Desiccant at 32° C. 50% RH.
D	High Temperature with a high range of humidities.	Water at 32° C. 50% RH.
E	High Temperature with a low humidity on one side and a high humidity on other.	Desiccant at 38° C. 90% RH.

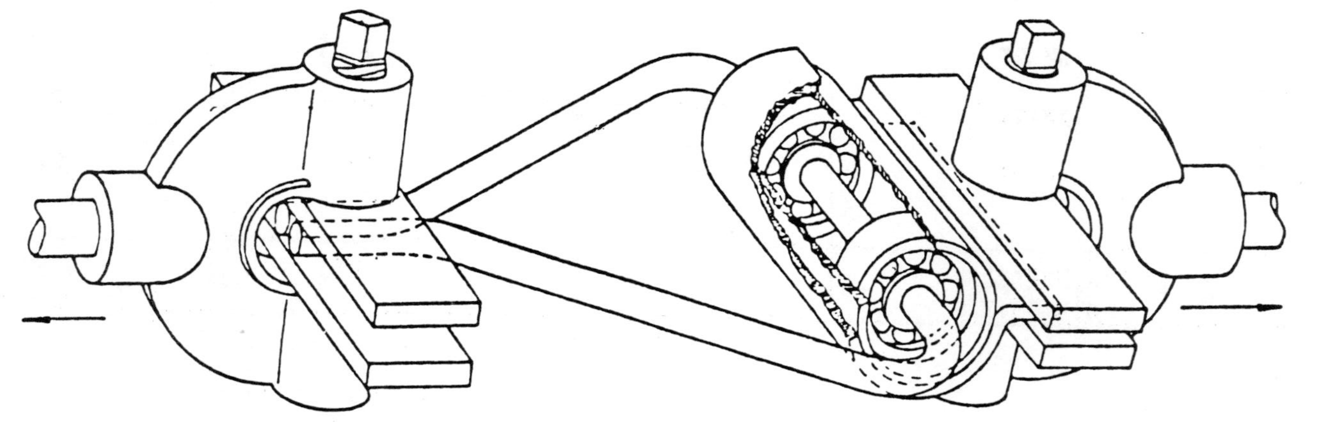

Fig. 4 90° Peel

Fig. 5 Chemical resistance PVC tape

REAGENT	IMMERSION TIME	TENSILE STRENGTH N/mm^2	ELONGATION %	DIELECTRIC kV	$M\Omega$ INSULATION RESISTANCE
Distilled Water	0	22	230	20	200×10^{12}
	2 Days	22	230	20	180×10^{12}
	7 Days	22	210	20	150×10^{12}
	21 Days	23	230	20	100×10^{12}
	5 Months	23	210	20	150×10^{12}
5 % Acetic Acid	2 Days	21	230	19	150×10^{12}
	7 Days	22	227	15	100×10^{12}
	21 Days	22	218	10	40×10^{12}
	5 Months	22	220	6	50×10^{6}
3% Sulphuric Acid	2 Days	21	227	20	200×10^{12}
	7 Days	23	212	20	150×10^{12}
	21 Days	24	210	20	150×10^{12}
	5 Months	23	220	20	170×10^{12}
2% Sodium Carbonate	2 Days	21	220	20	$[illegible] \times 10^{12}$
	7 Days	21	210	20	$[illegible] \times 10^{12}$
	21 Days	21	205	20	$[illegible] \times 10^{12}$
	5 Months	20	208	21	$[illegible] \times 10^{12}$
1% Sodium Hydroxide	2 Days	23	220	21	$[illegible] \times 10^{12}$
	7 Days	23	210	21	$[illegible] \times 10^{12}$
	21 Days	23	210	20	$[illegible] \times 10^{12}$
	5 Months	24	230	21	$[illegible] \times 10^{12}$

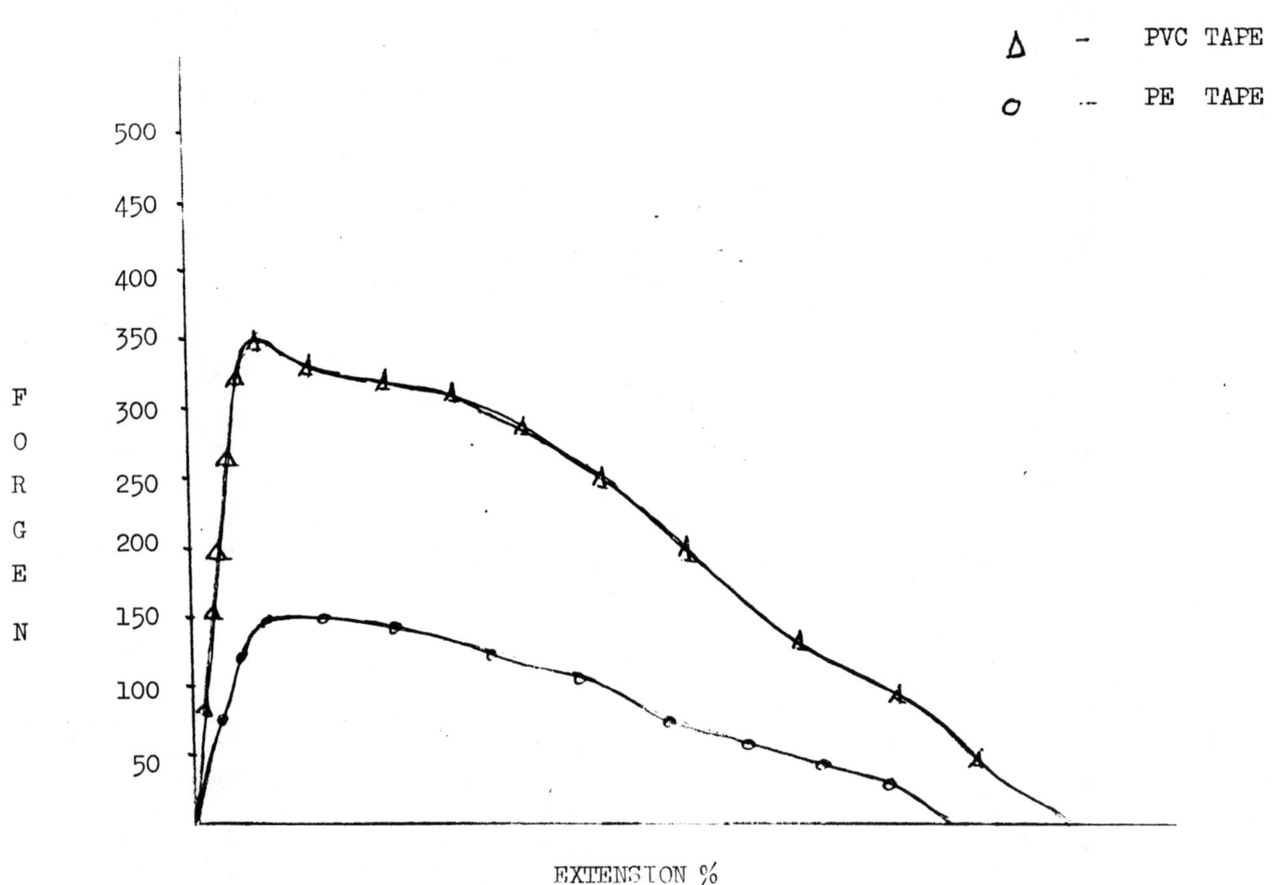

Fig. 6   Shear adhesion

internal and external
protection of pipes

September 5th - 7th, 1979

THIN FILM COATINGS FOR PIPES AND FITTINGS INTO THE 80's

P.A. White, ARIC. ARCATS

Mallatite Plastics Ltd., U.K.

Summary

Corrosion is not a new phenomenon. It is only recently, however, that the damage it causes has been fully recognised. The cost must not only be measured in the metal destroyed, but the enormous cost of replacement, maintenance and repair.

Whereas traditional and conventional anti-corrosion treatments merely limit the problem, and advancing technology of <u>Thin Film</u> thermoset and thermoplastic coatings goes further in the attempt to eliminate it.

The process involves the deposition of a Thin Plastic Film onto a metal surface. By choosing the correct polymer, the design engineer can combine the strength of the metal substrate with the decorative, insulative and anti-corrosive properties of the coating medium.

This Summer, the Thin Film plastic coating of pipes has quite dramatically become a requirement of the U.K. and European pipe and pipe fitting industry. Until quite recently, the preferred method of protection being high build bitumen, coal tar wrap and two pack systems. Typical Applications:

    i) OIL - B.P. on the Buchan and Magnus Fields.
        Thermoset – Fusion Bonded Epoxy (Tritorga 50 type).
   ii) GAS - British Gas, St. Fergus Line.
        Thermoset – Fusion Bonded Epoxy (Tritorga 50 type).
  iii) EFFLUENT - Bayer, Germany.
        Thermoplastic – Saponified EVA (Sintercoat 'E' type).
   iv) POTABLE WATER - VBB. Desalination Plant, Saudi Arabia.
        Thermoplastic – Saponified EVA (Sintercoat 'E' type).
    v) SLURRIES - I.C.I.
        Thermoplastic – Saponified EVA (Sintercoat 'E' type).

This paper discusses the back ground to corrosion and investigates up to date methods, materials and plant specially designed to reduce corrosion in the Pipe and Pipe Fitting Industries.

Held at Imperial College, London, England.
Organised and Sponsored by BHRA Fluid Engineering

CORROSION

1.1. INTRODUCTION

i) Corrosion is not a new phenomenon. It is only recently, however, that the damage it causes has been fully recognised. The cost must not only be measured in the metal destroyed, but the enormous cost of replacement, maintenance and repair.

1.2 WHAT IS CORROSION

i) The word, corrosion, is used for reactions between metallic materials and their environments in as far as these reactions damage the materials or cause them to undergo measurable changes. Corrosion and attempts to prevent it are old as the exploitation of metals by man. The last few centuries have been characterised by technical progress and the growth of industry.

ii) An enormous increase in the consumption of raw materials has raised the proportions of those ingredients in the atmosphere by which materials can be damaged; new technology and the extraction of raw materials from previously inaccessible and hostile environments, demand metallic systems that resist corrosion.

1.3 HOW IS CORROSION CAUSED

i) In nature, metals generally occur as ores in the form of oxides or salts of metals. They represent the states in which metals have their lowest energy contents and are therefore most stable. The extraction of metals from ores is accompanied by a considerable imput of energy. The state of a metal in its metallic form is therefore an unnatural one in which the metal has a high energy content. For this reason, all common metals have a tendency to recover their stable states, which are low in energy, i.e. to become salts, oxides or ores. Whether or not they are able to do so depends upon the ambient conditions. With certain metals, e.g. aluminium, copper and zinc and certain alloys, the formation of a patina (or dense oxide layer) on the surface, protects the lower layers of the material, thus preventing more corrosion than would economically be acceptable. These materials, which are said to be passivated, generally lose their metallic appearance as a result of this process. The great majority of metals and alloys, however, can only be protected from destruction by corrosion if suitable protective measures are taken.

ii) Corrosion has many forms, which include uniform surface corrosion, local corrosion, selective corrosion, dew point corrosion, microbiological corrosion, and the additional corrosion that is caused by mechanical stressing of the objects. Apart from reducing the useful life of an object, it increases the risk of failure while the object is in use. The aims of corrosion protection measures are, therefore, to delay the destructive action of a hostile environment, to extend the useful life of the object, and hence to save materials and replacement costs. Generally, these measures are divided into two categories.

1.4   <u>ACTIVE CORROSION PROTECTION</u>

   i) Active corrosion protection is carried out by influencing the
properties of the material, changing the reactivity of the
corrosive medium, or by applying electrochemical measures.

  ii) The properties of the metal can be changed by alloying it with
another metal. Stainless and acid resistant steels for example,
are produced in this way. The reactivity of the ambient medium
can be influenced by introducing inhibitors or removing such
stimulators of corrosion as oxygen. Anodic and cathodic
treatments are amongst the electrochemical measures that can be
taken to protect metals.

1.5   <u>PASSIVATING CORROSION PROTECTION</u>

   i) Passivating corrosion protection occurs when a barrier is formed
between the surface of the metal and the corrosive medium. The
medium is no longer able to attack the metal, because a protective
coating has been interposed between the two. If the protection
is to be really effective, the protecting layer must adhere well
to the surface of the metal, and it must be imperveous and
resistant to the medium.

  ii) The protective layer may consist of metal, e.g. chrome, nickel
or zinc, or an inorganic non metal, e.g. porcelain enamel, or of
such organic materials as rubber, paints, thermoplastic and
thermosets. The length of time for which a coating of an organic
material can protect an object increases roughly in proportion to
the film thickness.

SECTION A.2.

THE PROCESS

2.1   Two processes by which 250 – 500 micron thick films can be produced
from solventless, dry and free running powders were developed at
the beginning of the 1950's.

2.2   <u>THE ELECTROSTATIC PROCESS</u>

   i) The powders used for this purpose are mainly thermosets, with
particle diameters to a maximum of 80 micron.
<u>TRITORGA 50 TYPE</u>. (epoxy).

  ii) A pistol with a special head connected to a high voltage generator
gives the emitted powder a charge of up to 90Kv, causing it to
deposit a uniform coating onto the earthed and suitably pretreated
component. Preheating is only necessary when a thick coating is
required. The uniform coating is retained strongly enough to
withstand the handling involved in transference to the post fusion
and curing stage, which takes place at approximately 220°C.

2.3   <u>THE FLUID BED PROCESS</u>

   The powders used for this purpose are mainly thermoplastics,
with particle diameters ranging from (80 – 300) micron.
<u>SINTERCOAT 'E' TYPE</u>. (saponified EVA co-polymer).

The article, completely free of grease and rust, is preheated
in an oven to a temperature of between 200 and 250°C. It is then
immersed for a brief period in a tank containing aerated plastic
powder. Whilst in the fluidised powder, a uniform layer of
plastic fuses onto the surface of the hot component, and on
withdrawal, the fusion process carries on to completion.

SECTION A.3.

THE MARKET

3.1   OIL AND GAS TRANSMISSION (Service Temperature $\gg$80°C)

   i) Search for energy sources in previously inaccessible areas,
      has required new technology, no more so than in the North Sea.

  ii) Drilling for oil and gas has required a development and research
      programme comparable with the U.S.A. Space programme.

 iii) The hostile environment, the drilling depths, the high
      temperature of extracted oil, meant that previous technology was
      completely inadequate.

  iv) Standard forms of pipeline protection, coal tar wrap, bitumen,
      were completely unacceptable when considering the temperature
      of the recovered oil. Coal tar wrapped pipes began to move
      in their concrete jackets when subjected to the high oil
      flow temperature (120°C).

   v) Over the last three years, both British Gas and the Oil Companies
      have researched the problem. The outcome being the specification
      and acceptance of high build fast cure thermoset powders
      (Tritorga 50 type) applied by the 'electrostatic process' for the
      protection of 'pipelines' 'riser pipes' 'christmas trees' etc.,
      against corrosion.

3.2   WATER, EFFLUENT, SLURRY, GASEOUS TRANSMISSION (Service Temperature $\ll$80°C)

   i) Pipes for the transmission of Liquids and Gases such as slurries,
      effluent, potable water manufactured gas, can be made out of
      steel concrete or plastic. Invariably, if the pipe is of large
      diameter, the fittings are manufactured in steel, ductile, steel
      or cast iron.

  ii) To date, the standard form of finish was a bitumen dip, or for
      severe conditions, two pack epoxy.

 iii) However, to both match the improved performance of the concrete
      and plastic pipes, there is a growing demand for powder coated
      fittings, especially in the Middle East, where the humidity
      saline conditions and corrosive nature of the water, reduce the
      life expectancy of standard coated fittings to as little as
      18 months.

  iv) Recently great inroads have been made into the Middle East markets
      with Sintercoat 'E' on both large ductile iron pipes and fittings,
      and smaller cast iron pipes and fittings.

   v) A plastic pipe of 8" diameter costs approximately £25/m.
      A steel pipe of similar diameter costs approximately the same,
      without a protective finish.

vi) In the Middle East, if a steel pipe is used, it must be protected
by a good anti-corrosive coating, consequently plastic pipes
have been favoured because of cost. However, by nature of their
construction, the load bearing capacity is currently not as good
as steel pipe.

vii) Ductile iron and cast iron pipes cost approximately £10/m for an
8" diameter pipe, but because of surface profile could not be
effectively coated. Normally, this would not matter, as cast
iron is relatively inert; not so, however, in a Middle East
environment.

viii) The introduction of Sintercoat 'E', with its low melting point
and good flow, has overcome these problems and at a price of
£15/m on 8" diameter pipe, competes favourably with a plastic pipe.

ix) Where lower service temperatures are required, and a material that
is more flexible both in its physical nature and processing, then
the thermoplastic Sintercoat 'E', a saponified ethylene vinyl
acetate copolymer, applied by the 'Fluid Bed Process' is finding
increasing favour.

x) Sintercoat 'E', the latest and most sophisticated thermoplastic
produced to date, was developed eight years ago by Bayer, for
the 'Fluid Bed' protection of metals against corrosion.

SECTION A.4.
____

THE NEW TECHNOLOGY

4.1   FACTORY APPLIED THIN FILM COATING TO PIPES

Independent of whether pipes are to be laid on the land (land
lines) or on the sea bed (submarine lines), the powder coating is
applied to the pipe lengths (commonly known as joints) at the
factory.

The joints are then connected together on site, either using the
spigot and flange principal, or simply by welding. If welding is
used, approximately 3" of each joint length is left uncoated at
the factory.

4.2   FACTORY APPLIED THIN FILM COATING TO PIPE FITTINGS

To complement the system, tees, bends, valves etc., should also
be treated with a factory applied thin film coating, leaving
where necessary, the weld bevel free from coating for welding
purposes.

4.3   FIELD APPLIED THIN FILM COATING TO WELDED AREAS

4.3.1 LAND LINES

For Land Lines the joints are welded together prior to laying
in trenches.

4.3.2 SUBMARINE LINES

For Submarine Lines the joints are welded together according to
one of a number of methods

i) 'Drag Out' - here the joints are welded together on shore
and dragged out to sea.

ii) 'Lay Out' - here the joints are welded together on a
Lay Barge and laid out at sea.

iii) 'Reel Out' - here the joints are welded together on shore,
reeled onto a Catherine wheel, placed on a
Reel Barge, and reeled out at sea.

4.3.3 THIN FILM COATING

After welding, the weld areas should ideally be coated with a
system compatible with the factory applied coating. Hence
there is a requirement for a technique to apply a Thin Film
Coating of a Fusion Bonded Epoxy or Thermoplastic Sintercoat 'E'
to the weld area.

SECTION A.5.

THE TECHNIQUES AND PLANT FOR COATING

5.1   INTRODUCTION

i) Mallatite Plastics, a leading company in the anti-corrosion field,
aware of the rapidly emerging requirement for Thin Film Coatings,
has commissioned, this Summer, with the assistance of interested
parties in the Oil, Gas and Water Industries, a unique range
of Thin Film Coating Plants.

ii) These facilities will complement the high throughput pipe coating
plants of British Gas, Key & Krammer, and Blome, and will provide,
in addition, an anti-corrosion pipe fitting service to those
manufacturers producing pipe systems, whether they be plastic,
cast, malleable etc.

5.2   OIL AND GAS TRANSMISSION (Service Temperature $\gg 80^{\circ}$C)

5.2.1 PIPES

i) Fusion bonded epoxy coatings are applied to pipes using an
automatic process.

ii) In Europe, there are only two plants that produce significant
amounts of Thin Film Plastic coated pipes,

Key & Krammer in Holland - maximum processing diameter 12".

Blome in Germany - maximum processing diameter 42".

and more recently the British Pipe Coating Plant at Leith,
commissioned, the Summer of this year - maximum processing
diameter 42".

iii) After shotblasting to SA $2\frac{1}{2}$, on a poleabrator, the pipe is
preheated using an indirect heat source to a temperature of
approximately 250°C. The heating is followed by the automatic
application of electrostatically applied Fusion Bonded Epoxy
coating, to a thickness of approximately 450 micron, the coating
being water quenched after curing.

iv) Each of these plants has a capacity of approximately 10KM/week
    of 4" diameter pipe, being designed to handle nothing but
    straight lengths.

5.2.2 <u>PIPE FITTINGS</u>

i) Pipe fittings are pretreated in a similar manner, but because of
   their geometry and wall thickness, are sprayed by hand in an
   enclosed booth using a powerful electrostatic or flock gun with
   an output of 1 - 2 Kg/min.

ii) One of the first plants in Europe built specifically for this
    technique was installed at Mallatite in April, 1979.

5.2.3 <u>WELD AREAS</u>

i) All rust, scale and dirt is first removed from the bare steel
   around the weld area, and about 2" of the factory applied coating
   by grit blasting.

ii) The weld and roughened coated area is then heated using a hinged
    heating coil, the temperature of the pipe being raised to a
    temperature of 260ºC. The temperature is monitored and recorded
    on a chart recorder.

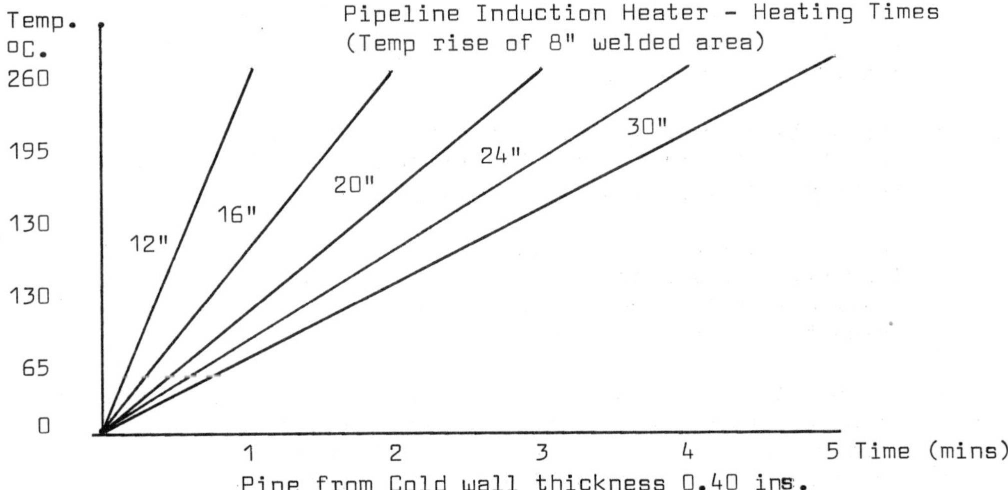

iii) For pipe diameters up to 12", the powder is applied by the fluid
     bed method; diameters above 12" by flock spraying.

iv) The powder being very reactive, gels and cures within 1 min.
    A temperature vs time curve is plotted and recorded to establish
    full cure; after cooling, the coating is tested for holidays.
    Repairs are carried out as necessary using two pack epoxy touch up.

5.3 <u>WATER EFFLUENT, SLURRY, GASEOUS TRANSMISSION</u>
       (service temperature ≪ 80ºC)

5.3.1 <u>PIPES</u>

i) Sintercoat 'E' is used as a coating on suitably pretreated pipes
   where the service temperature is less than 80ºC, or where epoxies
   cannot be used because of their processing limitations.

   e.g. a) Pipes with poor surface finish.

         b) Cast Iron Pipes.

         c) Ductile Iron Pipes.

ii) The technique that has evolved is demonstrated on the K9 unit
at Mallatite. Essentially the external portions are sprayed
automatically by a travelling robot, whilst the internals are
rotationally coated.

5.3.2 <u>PIPE FITTINGS</u>

i) Once again, the technology is new; Pretreated fittings weighing
up to $3\frac{1}{2}$ tonne are preheated to a temperature of $250^{\circ}C$ prior to
being immersed into a sunken fluid bed containing 10 tonne of
Sintercoat 'E'.

ii) The crane used for the lifting and dipping operation has a lift
speed of 70'/min.

This technique ensures that a uniform thin film coating is applied
simultaneously to the entire surface of the fitting.

iii) Repairs are carried out using the 'flame spray gun', a soldering
torch, or alternatively using a 'two pack' touch up system.

5.3.3 <u>WELD AREAS</u>

i) Fusion Bonded Epoxy coatings, unlike the thermoplastic
Sintercoat 'E', have to be fully cured before their full physical
and chemical properties can be realised. As the cure is a
function of temperature and time, it is essential that graph
recordings are taken.

ii) Thermoplastics, however, by nature assume all their properties
when the material flows. The flowing process is visible and
therefore a temp/time graph would be superfluous.

iii) The technique of applying, to the suitably prepared weld area, a
thin film coating of a thermoplastic, which forms a homogenous
coating with the factory applied finish, has now been established
using the Flame Spray Gun.

iv) In the Flame Spray Gun, a flame is produced using oxygen and
propane. This flame is used to preheat the metal surface and
factory applied coating to approximately $150^{\circ}C$. The powder
contained in a hopper is then air fed through the centre of the
flame; it melts, and is projected onto the weld area and
roughened factory applied coating, producing a homogenous and
permanent repair.

SECTION A.6.

6.1 <u>CONCLUSION</u>

As raw materials become more expensive, either because of
their scarcity, or because of the greater cost in extracting
them from more inaccessible regions, the technology must be
to thin metal sections, and to <u>Thin Film Coatings.</u>

<u>Thin Film Coatings</u> in the 80's must, therefore, provide an
anti-corrosion finish, equal to, if not better than, the
thick film coatings of the 70's.

internal and external
protection of pipes

September 5th - 7th, 1979

FUSION BONDED EPOXY PIPELINE COATINGS – A REVIEW

R.F. Strobel, P.E.

3M Company, U.S.A.

and

B.C. Goff, Grad. P.R.I.

3M (U.K.) Limited

Summary

The increasing need for liquid and gaseous energy and the current fuel shortages has initiated a worldwide search unequalled in past history. Transport of these fuels from geographically remote sources to areas of use is economically and safely accomplished using pipelines. Pipelines are subjected to hostile environmental conditions during transportation, installation and use and must be protected from these adverse conditions, this being most easily accomplished through the use of coatings.

The pipeline coating industry throughout the world has grown considerably over the past ten years, especially in Europe with the development of the North Sea oil and gas industries, and further by the increased demand for pipe to be coated in Europe and exported to the Middle East.

It is accepted that the conditions of the environment in, for example the North Sea or Lake Maracaibo in Venezuela are of the most aggressive in the world, and it is essential that pipeline coatings have protective properties of the highest order. Conventional coating systems have been seen to fail due to these more aggressive environments and also with the development of more modern pipe laying techniques has meant new systems have been designed to cope with these extra demands.

One such system is the fusion bonded epoxy coating developed by 3M Company in 1960. These coatings are now gaining acceptance in the world-wide oil, gas and water industries as being superior to conventional systems where increased physical and chemical properties are necessary.

Held at Imperial College, London, England.
Organised and Sponsored by BHRA Fluid Engineering
Copyright BHRA Fluid Engineering, Cranfield, Bedford, England.

1.0.   <u>Discussion</u>

During the past ten years the pipeline industry
has been experiencing a gradual but definite
transition away from the widely used bituminous
and coal tar type coatings, particularly on large
diameter pipe. Reasons are varied but problems
with the application and installation of bituminous
coatings seem to be the basic reason for the trend.
Bituminous coatings are applied in fixed plants
and over the ditch. Over the ditch coating, though
most economical, suffers from poor surface
preparation and quality control, whilst the
application of the coating can also be subjected
to inclement weather delays. Precoated pipe
provided by fixed plants is difficult to ship,
handle, bend and install without damage to the
bituminous coating, particularly in cold weather.
The bituminous coatings may also flow and sag, both
prior to, and following installation, whilst damage
to the coating can result from soil stress action
and backfill compaction.

Continued research on pipeline corrosion and stress
corrosion cracking, along with active concern by
governmental agencies and environmental groups over
pipeline safety and the impact of pipelines, has
forced a new look at corrosion control. The high
cost of pipeline construction and maintenance, and
the need to transport a wide variety of liquid and
gaseous fuels at varying temperatures from remote
areas in the world, has created unusual construction
and service conditions that can only be met by more
sophisticated coatings. At present fusion bonded
epoxy coatings most nearly meet these needs. Reasons
for the increased use of fusion bonded epoxy powder
to coat pipelines in Europe can be explained as
follows

1.1.   <u>Increased Environmental Conditions</u>

Pipelines transport large quantities of liquids and
gases, and are frequently pressurised. When these
liquids or gases are pressurised, heat is developed,
which can adversely affect the organic coating.
Thermoplastic materials, polyethylenes, polyvinyl
chlorides and bitumous coatings will soften, with
subsequent loss of adhesion to the steel pipe surface.

The pressure in the pipeline causes the pipe
to flex or swell. This could produce cracking
in conventional systems especially when pressure
testing pipe during commissioning. In the North
Sea, for example in the Buchan or Magnus fields,
oil is being or will be pumped under pressure
resulting in oil temperatures approaching 220°F.

Oil companies have instigated a research
programme to seek an alternative system that can
stand up to these higher temperature demands.
Fusion bonded epoxy coatings were found to be
able to cope with these temperature and pressure
requirements.

1.2.   Methods of Laying Pipe

1.2.1   On-Shore

Conventional methods of laying pipe can involve
bending the pipe to counter the natural terrain
of the ground. Fusion bonded epoxy coatings are
designed to meet the bending requirements of the
American National Standards Institute ANSI B31.4
and B31.8 standards without cracking, crazing
or disbondment. High durometer polyurethane
padding on the stiff back, bending die and pin up
ring of a standard field bending machine prevents
damage to the coating. Field bends on the epoxy
coated pipe are easily made with the padded
machine and the addition of padding does not
increase the tendency towards wrinkle bends. The
coating is also unaffected by natural soil chemicals
and is particularly resistant to the mechanical
action of soil stress and backfill compaction.

1.2.2   Off-Shore

The most recent advance in laying pipe offshore
has come with the introduction of the reel barge.
Operated by Santa Fe, these barges have a central
reel, upon which can be wound up to 50 miles of
pipe, depending on the diameter. The barge is
then shipped out to wherever the pipe is to be
laid, the pipe unreeled, straightened and then
conventionally laid.

Individual lengths of pipe are first coated
with fusion bonded epoxy powder, a 2 inch
cut back being left at the end of the pipe for
welding purposes. The pipes are then shipped
to the dockside and welded together into
approximately 500 metre strings in preparation
for spooling aboard the reel barge. Girth welds
are protected by fusion bonded epoxy powder,
the joints being initially grit blasted to near
white metal, then heated with an induction
heater to approximately $450^{\circ}F$. Epoxy powder
is then sprayed on the joints to a thickness the
same as the main pipe sections. All this work
is carried out on site at the dockside prior to
spooling. The long pipe strings are then
consecutively spooled onto the reel - as the
back end of one approaches the barge, a new
string is welded on to it to continue the line.
Pipe used is heavy wall to obtain negative
buoyancy, with coating thicknesses averaging 16
thousandths of an inch.   The coating is subjected
to a variety of stresses and strains, especially
high compression loads during the spooling of the
pipe on and off the barge. When reeling off, it
has to go through a series of straightening devices
where the pipe is reverse bent. Many miles of
fusion bonded epoxy coated pipe have been laid by
this method in the Gulf of Mexico and British
Petroleum are pioneering the laying of pipe in
the North Sea by this method in the Buchan field
during the summer of 1979.

Fusion bonded epoxy coated pipe can also be laid
conventionally off a lay-barge, and due to its
high impact and abrasion resistance properties
can resist damage during the laying process off
the barge and, most important, can resist
abrasion and impact damage when in contact with
the sea bed.

1.2.3. <u>Field Welds/Damage Repair/Shipping</u>

Coating field welds on site has always been a
problem. Tape wrapping or shrink sleeve systems
have provided an adequate level of corrosion
protection but are time consuming and usually do
not possess the same properties as the parent
system on the main pipe.

Two pack repair systems, although easily
applied, suffer from time delays due to
cure time. It is ideal if the field welds
are coated with the same system as the main
pipe, and this is when the induction heating
and powder spraying technique previously
described is used. Large areas of mechanical
damage may be repaired in this manner, but
pinholes and other small areas are repaired
using a hot-melt patch stick. This stick of
a solid thermoplastic material has been
designed with time as one consideration. For
example, on a lay barge, if a pinhole is
discovered just prior to the pipe entering the
water, one may have only a few minutes to effect
a repair. The area to be repaired is first
cleaned and then roughened to ensure good
adhesion. Using a non-contaminating heat source
the area is heated to approximately $350^{\circ}F$ in a
manner which avoids burning or charring of the
epoxy coating. Whilst continuing to heat the
cleaned and preheated area, the patch compound
is applied by rubbing the stick on the repair
area in a circular motion to achieve a smooth
neat appearing patch having a thickness of
approximately 15 thou.

This method of repair is currently being used
by the British Gas Corporation in the U.K. for
pinhole repairs on all their fusion bonded
epoxy coated pipe.

Fusion bonded epoxy coatings are extremely tough
and resistant to handling damage, however, as
with any coated pipe care must be taken to
properly handle, load and transport the coated
pipe. Shipments have been made for thousands of
miles by truck, rail and barge and the coated
pipe arrives at the job-site in excellent condition. As an example, 42 inch O.D. pipe was double
-jointed into 80 foot (24 metre) lengths and
coated at a plant in the U.S. with fusion bonded
epoxy powder. The coated pipe was then loaded on
rail cars, shipped 3000 miles, unloaded into stock-
pile, then strung along the ditch and welded up.

This represented nine coated pipe handling
operations from the coating plant to the
ditch. The final electrical inspection
detected less than 1.5 'holidays or jeeps'
per joint, a very low number.

2.0.   <u>Development of Pipe Coating Plants in Europe
to Specifically Handle Fusion Bonded Epoxy Powder</u>

There are currently three major pipe coating
plants in Europe that can successfully coat with
fusion bonded epoxy powder resin. All three
operate a similar type of plant, using a 'spiral
in-line' process of pipe coating. This involves
the pipe moving forward through the grit blasting
unit, oven and spray coating head in a spiral
fashion, as opposed to a 'transport' type of
plant which involves a moving coating head unit
traversing a static pipe.

The Key and Kramer plant in Holland can handle
pipe up to 12 inches in diameter. Grit blasting
is carried out automatically using a wheel
abrader system, and heating is by courtesy of an
indirect gas fired batch oven. So far this year
work has been concentrated on coating flow lines
carrying hot oil for major oil companies including
Shell, BP and Phillips Petroleum.

The Anton Blome plant in Germany can handle pipe
up to 56 inches in diameter on their main line,
and due to increasing business are currently
constructing a second line designed to coat small
diameter pipe. They are by far the major user
of fusion bonded epoxy powder in Europe, carrying
out projects for Shell in the Middle East, and
other oil companies' pipe destined for the North
Sea. Again, this plant is built on the 'spiral
in-line' principal, utilising a direct fired gas
oven feed by pipe coming from a wheel abrader
type grit blasting unit.

The British Pipecoaters plant at Leith in
Scotland has recently been commissioned and can
handle large diameter pipe up to 60 inches on
one line, and small diameter pipe down to 4 inches
on the other line.   Both these lines are fed from
a common wheel abrader blasting unit and then
through to the specific induction heating coil.
They are carrying out a large contract for British
Gas Corporation coating gas transmission pipe
(42 inches in diameter) and are supplementing this
work with contracts from oil companies.

3.0.   <u>Development of Custom Coaters in the U.K to Coat
with Fusion Bonded Epoxy Powder</u>

The custom coating industry has grown in the U.K
during the past year to deal with two major
aspects of the pipe coating industry:

(a) To coat ancilliary equipment for pipe lines
    e.g. bends, valves, T-pieces etc.

(b) To coat short lengths of pipe, mainly in the
    form of fittings for the water industry.

Obviously, all this ancilliary equipment cannot be
coated in a line pipe coating plant due to
configuration and design. Applicators have therefore been developed to specifically coat these
articles either by fluidised bed powder application
or electrostatic/flock spray powder application.

3.1. Methods

The fluidised bed consists of two chambers
separated by a specially designed porous membrane
which serves to diffuse air uniformly throughout
the coating powder. In proper operation, the resin
expands to twice its original volume ready to
accept preheated objects. This method is probably
the fastest way of coating large articles but care
has to be taken over the temperature of the metal
article prior to dipping, and the length of time
the article is in the bed. Once this relationship
has been established, large metal articles of sizes
up to, for example, 2 metres x 2 metres x 3 metres
can be successfully dip coated.

Electrostatic coating is accomplished by charging
powder particles with a high voltage and spraying
onto an object at ground potential.

With the proper powder recovery system, the
overspray powder can be recirculated for
utilisation of up to 98% of the coating material.
Coating by this method is ideal for flat, angular
or irregularly shaped objects or those which
cannot fit into a fluidised bed.

Coating by flocking requires the least equipment
and can be used to maximum advantage where deep
recesses in the object to be coated necessitate
a forced air application. Basically, all this
method involves is blowing a powder and air
mixture directly onto a pre-heated surface.

With all these techniques of coating, it is
important that the surface of the steel fitting
is properly prepared to a specification issued by
the manufacturer. This usually entails inspecting
the surface prior to grit-blasting and pre-cleaning
to remove any traces of oil, grease or loosely
adhering deposits. The metal is then blast cleaned
to BS4232:1967 Second Quality of Surface Finish
(Swedish Standard SIS 055900 S.A.2.5) having a
surface profile of 40-75 microns.

4.0. Project Work and Current Pipeline Coating Contracts Relevant to Europe

4.1. Oil Industry

Shell and BP are the first U.K. based companies to
use fusion bonded epoxy powder on line pipe which
is being coated in Holland and is destined for the
North Sea. These flow lines are carrying oil at
temperatures up to 220°F and the companies carried

out a joint research programme, eventually deciding
upon fusion bonded epoxy powder. Phillips
Petroleum have had 32 kilometres of small bore pipe
coated in Holland for their North Hewitt field with
fusion bonded epoxy, and Chevron opted for fusion
bonded epoxy coating on 6 kilometres of 8" diameter
pipe for the Ninian field. Conoco are coating 5
kilometres of 12" pipe in Germany with fusion
bonded epoxy and have a further 19 kilometres of
16" diameter pipe awaiting coating; all pipe
destined for the Murchison field. Both Texaco and
Amoco are considering the use of fusion bonded
epoxy coating for their high temperature oil flow
lines, whilst Shell have specified fusion bonded
epoxy for an on-shore gas transmission line.
Shell have also opted for fusion bonded epoxy
coating on 450 kilometres of 18" diameter
pipe destined for the Middle East - work to
be carried out in Germany.

4.2. Gas Industry

As a result of a large research programme carried
out over the last three to four years, British
Gas are to use fusion bonded epoxy coating on all
their future gas transmission line. The first
phase of the fourth Scottish feeder is now
approved, and 450 kilometres of 42" diameter pipe
will now be coated with fusion bonded epoxy
instead of the usual coal tar preparation. Work
will be carried out at the British Pipecoaters
plant in Scotland.

Interest in fusion bonded epoxy coating has also
been shown by the gas industry on the Continent.

4.3. Water Industry

The Water Boards in the U.K. are now looking at
fusion bonded epoxy coating as an alternative to
the usual nylon or PVC coatings. The added
advantages of epoxy in terms of its physical and
chemical properties, together with the ease of
application, has increased the business in terms
of fusion bonded epoxy coated fittings over the
past year. The National Water Council have
approved the product as suitable for use with
potable water. Most of the work being carried out
on water pipes and fittings in the U.K. at present
is for export to the Middle East. The applicator
in question, located in Birmingham, is fabricating
the pipes and fittings and then coating them via
fluidised bed or flock spray. Pipes handled can
be up to 2 metres x 2 metres x 3 metres in size.

5.0. Project Work and Current Pipeline Coating Contracts
      relevant to the United States

Almost every major oil, gas and pipeline company
in the U.S. is using fusion bonded epoxy coatings
for corrosion protection of their pipelines. The
proposed Sohio pipeline, which is designed to move
Alaskan crude oil from California to the Middle
and Eastern parts of the U.S. have specified
fusion bonded epoxy coatings. The U.S. Federal

Energy Administration used fusion bonded epoxy coatings on a 160 kilometre large diameter crude oil supply line in Southern Louisiana whilst Tennessee Gas Pipeline Company used these materials internally and externally on several kilometres of a large diameter off-shore gas line. This line is designed to move the wet gas from the off-shore platform to land based terminals. The pipeline joints were also coated internally and externally with fusion bonded epoxy powder to provide complete corrosion protection for this very critical line. Tennessee Gas was instrumental in the development of the concept of coating girth welds internally and externally with epoxy powder coatings, this interest resulting from superior performance of these materials in their extensive laboratory testing programme and 'in-the-field' experiences .

Fusion bonded epoxy coatings have been used for several years to protect fittings and water systems. However, their use has been restricted because there was no standard approved by the American Water Works Association. This was corrected in March, 1979 with the approval of Standard C213-79.

6.0. <u>Project Work and Current Pipeline Coating Contracts relevant to Latin America</u>

Venezuela has been the leading country in Latin America in the acceptance of fusion bonded epoxy coatings. The first pipe was coated in 1971 for installation in the corrosive water of Lake Maracaibo where other coatings have failed to perform effectively, indeed this lake is considered to be one of the most destructive environments for pipeline coatings. Some of the first pipe coated was inspected in early 1979 and the coating found to be completely intact and exhibiting physical properties similar to those at the time of installation.

The use of fusion bonded epoxy coatings has spread over the past eight years to include government owned gas and water companies and essentially every segment of the parent Petroven Oil complex.

7.0. <u>Conclusion</u>

Epoxy fusion bonded coatings represent a significant improvement in pipeline coating technology. Sufficient field experience has been gained over the past decade to substantiate the advantages that accrue to the pipeline owner. Although the United States have been leaders in the field of application of fusion bonded epoxy coatings for many years, epoxy powder coating plants and custom applicators are now operating successfully in Europe.

Fusion bonded epoxy coatings are now being specified by the oil, gas and water industries throughout the world.

8.0. <u>Acknowledgements</u>

Case history information from the United States provided by courtesy of 3M Company, St. Paul, Minnesota, U.S.A.

Case history information from the United Kingdom provided by courtesy of 3M (U.K.) Limited, Bracknell, Berkshire, U.K.

internal and external
protection of pipes

September 5th - 7th, 1979

RECENT DEVELOPMENTS IN MATERIALS & METHODS OF APPLICATION OF POLYETHYLENE PIPELINE COATINGS

G.M. Harris

The Kendall Co., U.S.A.

Summary

During the past few years the field of polyethylene pipeline coating tapes has been broadened by the new developments in products and application techniques. These new developments are discussed in terms of:
1) The use of sophisticated scientific instruments to study the rheological and thermal properties of the major raw material and compounds used in these products.
2) A description of three new products.
3) The application and performance experience of polyethylene tape coatings on pipelines built under Arctic-like winter weather conditions.
4) The correlation of laboratory test results and the in-service performance of polyethylene tape in terms of coating resistance.
The conclusion is drawn that the continued growth in the use of polyethylene tapes as pipeline coatings is due to their proven record of successful performance over the past thirty years. This in-service success is due to the inherent properties of the raw materials used in the tapes and the scientific basis for understanding how to use these properties to the best advantage.

Held at Imperial College, London, England.
Organised and Sponsored by BHRA Fluid Engineering

Pipeline coatings products can be characterized by their major raw material components. At the present state of underground pipeline coatings technology, there are three raw materials which dominate the field (Ref.1). This being the case, corrosion engineers responsible for mitigating the external corrosion of underground pipelines most frequently select coatings from the three raw material areas of coal tar, polyethylene or epoxy. Although none of these three materials are recent developments in themselves there have been recent products and application advancements made within these three categories. A significant number of the new advances in coating technology are found in the polyethylene category. There are some within the extruded polyethylene coatings group and others within the group called polyethylene tapes. It is the purpose of this paper to delineate the recent developments in materials and methods of application of polyethylene tape pipeline coatings.

BASIC CONSIDERATIONS IN DESIGNING POLYETHYLENE PIPELINE COATING TAPE

In designing Polyethylene Pipeline Coating Tapes, a number of basic scientific tools can usefully be applied. Some significant examples involving Rheometers and Differential Scanning Calorimeters are given below. (Ref. 2)

Adhesive Design

The adhesives on pipeline coating tapes must be able to conform to a substrate readily to establish good and permanent contact. In the case of a steel pipe, the substrate is relatively rough, and application rates quite high. The adhesive mass must show quick bonding characteristics when first applied and then high resistance to creep for long periods and, in some cases at relatively high temperatures.

Quantitatively, in rheological terms, the measure of conformability is shear compliance, which is the amount of shear deformation obtained per unit load for unit cross-sectional area. Shear compliance is a fundamental rheological property of a material. In the case of adhesives, as with other polymeric substances, it is a viscoelastic property, as an adhesive must have both viscous and elastic characteristics so that it can bond quickly, flow, and finally resist creep movements for long periods.

Viscous and elastic characteristics can be separately determined. A standard way of doing this is to apply a sinusoidally-varying stress to the adhesive being examined, and to measure the in-phase and out-of-phase components of the resulting strain. Instruments called Rheometers are available for this purpose. A "Rheovibron" is a Rheometer which induces very small strains (less than 1%) in an adhesive at frequencies of 3.5, 11, 35 and 110 Hz., over any desired temperature range from about $-150^{\circ}C$ to $200^{\circ}C$. Data obtained from a "Rheovibron" can be used to calculate, for example, the elastic (storage) compliance, $\underline{J}'$, and the viscous (loss) compliance $\underline{J}''$.

A pipeline tape adhesive consists of an elastomer and a tackifying resin. Addition of tackifier to the base elastomer of an adhesive has the following effects:

1) It raises the compliance in the "rubber plateau" region, permitting easier flow of the adhesive.

2) Above a certain concentration of tackifying resin a two-phase system may result: resin dissolved in rubber (continuous phase) and low-molecular-weight rubber dissolved in resin (discontinuous phase). This

two-phase nature appears to contribute to quick bend
development on the steel substrate.

3) The glass transition temperature (Tg), below which the
system becomes a brittle, resinous plastic, shifts to
higher temperatures.

4) The compliance at low temperatures rises. This change
may leave inadequate cohesive strength - creep resistance,
etc. - and crosslinking may be necessary to overcome this
deficiency.

As (3) and (4) occur, the rubber plateau - which is responsible
for what is termed "rubber elasticity" - gradually shrinks until,
at some resin level, it vanishes completely. Rubbery behavior results
from entanglements of the polymer chains. When a stress is applied
to the adhesive system, the chains cannot readily disentangle, and
are instead stretched out. In this extended condition they exert an
elastic restoring force. In order to obtain such entanglements,
materials must have a certain minimum molecular weight, (M_e) below
which rubbery behavior is not observed. As tackifier content increases the elastomer concentration, is diluted, and a higher M_e would
be needed to provide entanglement coupling than for the plain elastomer. Alternatively, the tackifier can be considered as reducing the
average molecular weight of the system. If "permanent entanglements"
such as, interchain crosslinks between molecules are put into the
elastomer, the rubbery behavior can be maintained at much higher
temperatures, and for longer times.

Figure 1 gives an example of typical changes in the loss compliance of an adhesive as tackifier is added in greater amounts to an
elastomer. The glass transition temperature shifts by about four
decades of frequency as the tackifier content goes from 0% to 80%.
This change is accomplished by a rise of almost 100-fold in the
viscous compliance J".

As an example of a direct application of such experiments,
Figure 2 shows the compliance of two experimental adhesives when
different tackifiers were used, measured at a frequency of 1 Hz. In
the case of the composition A, glassy behavior begins in the vicinity
of $-10^{\circ}C$. The other adhesive, B, however, begins to become glassy at
about $25^{\circ}C$. It was predicted from this data that the former would
perform satisfactorily under normal use conditions, while the latter
would not. This was amply borne out in actual use tests, where the
second adhesive tended toward brittle failure near room temperature
$(22^{\circ}C)$.

Figure 3 show the effects of crosslinking on an adhesive system,
based on an elastomer with a molecular weight of 80,000. Relatively
little change is found at high frequencies, indicating good conformability, but the compliance as the amount of crosslinking increased,
is drastically reduced at the low-frequency end which corresponds to
high temperatures.

Measurements of viscoelastic properties of adhesives on tapes can
also reveal a great deal about the effects of aging, where time,
temperature, and environmental conditions (oxygen, for example) all
play their parts. As an example of this, Figure 4 shows the changes
which two adhesives underwent on accelerated aging for 500 hours at
$100^{\circ}C$ with access to air only by diffusion through the polyethylene
backings. One adhesive is made from butyl rubber and the other from
natural rubber. Only relatively modest changes were found in the
butyl-based adhesive - some increase in high-temperature compliance.
The natural rubber adhesive became stiffer, the compliance dropped

throughout, and the glass transition shifted to a higher temperature. These results show that butyl adhesives are stable for long periods on underground pipes while natural rubber adhesives are unstable and lead to coating failure in a relatively short time.

Backing

Polyethylene backings do not normally show very serious changes with aging, unless very drastic conditions exist. However, with the increasing demands on pipeline coatings for long-term, high-temperature stability, it is important that polyethylene tapes are able to function properly under rather severe conditions.

If a tape is made only from high-density polyethylene, the tape would have greater strength at higher temperatures. Unfortunately, the very high crystallinity of this material leads to a high modulus and, therefore, a low compliance. This makes the tape too stiff to be wrapped around a pipe in a layer thick enough to give good protection. Therefore, to obtain a flexible enough tape backing, a lower-density PE, or blends of high and low PE must be used. Figure 5 shows the compliance of a 100% LDPE tape, a 100% HDPE tape, and a tape made from a blend of the two. The first is a serviceable film; unfortunately, it melts at about 110°C. The HDPE tape does not melt until 130°C but is too stiff for easy applications. If a blend of the two is used, minor melting occurs at 110°C but final melting (and consequent total loss of strength) does not take place until almost 130°C and thus a useful amount of strength is retained up to 125°C. The polyethylene pipeline coating tape with the greatest experience on underground pipelines has a backing consisting of a blend of high and low density polyethylenes. This product has the optimum application and in-service properties.

A practical instance of this behavior can be found in Figure 6, where three commercial pipeline coating tape backings are compared. One is based on plasticized PVC, and it is obvious that its aging characteristics are thoroughly unsatisfactory because the plasticizer is lost, and the PVC tape backing becomes stiff and brittle. The other two are polyethylene tapes, although one contains an "anchoring layer" which is not polyethylene. The curve of the unaged specimens shows that the non-polyethylene anchor layer melts about 15°C below the pure polyethylene layer and thus this tape has a lower service temperature than the tape made only with polyethylene.

In practice, the results are more dramatic. Both materials were aged for periods of 500 hours in boiling water. The tape based on a blend of LD/HD polyethylene was virtually unaffected even after 500 hours exposure while the tape with the anchor layer of a non-polyethylene resin failed completely after about 200 hours because complete delamination of backing from adhesive took place.

The reason for this failure by delamination was not immediately obvious, and therefore, the two tapes were examined by a different scientific analytical technique.

Differential Scanning Calorimetry curves were made of both tapes in order to observe their melting behavior (Figure 7). The PE only backing tape showed the expected results – melting of the LDPE component at about 110°C, and the HDPE near 130°C. The other backing showed characteristic melting of a medium density polyethylene at 124°C but it had another melting peak at about 95°C. This second melting peak at 95°C was below the temperature of the boiling water used in the aging test. It turned out that an ethylene/vinyl acetate copolymer was used as the "anchoring layer", between adhesive and backing. This gave good results at low temperatures, but when exposure temperatures exceeded 95°C melting point of the anchoring layer, catastrophic failure resulted.

NEW PRODUCTS

Three new products in the category of polyethylene tape pipeline coatings will be discussed herein. These are:

1. Primers with stress corrosion cracking inhibitive properties.

2. Primer type coatings for the protection of shot blast cleaned pipe during a storage period bridging the time period between the cleaning operation and the over-the-ditch construction and tape coating of the pipeline.

3. Butyl rubber based adhesives which are crosslinked in order to improve their bonding qualities at elevated temperatures.

Primers with Stress Corrosion Cracking Inhibitors

Several years ago the American Gas Association contracted The Battelle Institute to do research in the area of the stress corrosion cracking of high pressure gas pipelines. This research led to a better understanding of the stress corrosion cracking (S.C.C.) process and to the discovery of several inorganic salts which will inhibit the S.C.C. process when properly introduced in to the environments which promote S.C.C. (Ref. 3 and 4).

Concurrently Battelle also developed the test methods whereby the S.C.C. mechanisms and the effectiveness of the newly discovered inhibitors could be studied (Ref. 5).

The American Gas Association through the scientists at The Battelle Institute told the manufacturers of pipeline coatings about the research studies which led to the discovery of the inorganic salt inhibitors. The scientists suggested that these chemicals could be incorporated in some components of the various coatings as a means of bringing the discovery of the inhibitors to a practical end use. The Research and Development divisions of polyethylene tape manufacturers have been able to incorporate the AGA/Battelle S.C.C. inhibitors in the tapes primers and to demonstrate their inhibitive capabilities.

The proof lies in the results obtained from testing the primers containing the inhibitors by the methods and techniques developed by The Battelle Institute. These test methods and techniques have been described in papers presented by Berry, W.E., and others so it is not necessary to describe them here (Ref. 6). The test results for a polyethylene tape primer containing sodium tripolyphosphate are given in Table 1. At the present time the primer whose test results are listed in Table 1 is in commercial production and is being used in very large quantities on pipeline throughout the world.

Primer Coatings to Protect Cleaned Pipe During Storage

The planning of very large pipeline projects often involves the manufacturing and delivery of the pipe well in advance of the actual construction dates. This can lead to the storage of the pipe at strategic locations along the pipeline right-of-way as well as long transport times. Experience has shown that unprotected pipe during transit or storage out of doors can experience considerable surface corrosion before the pipe is coated before it is buried. While the surface corrosion may result in the loss of significant amounts of

steel, it may also present a rough surface contour which may cause
problems when the pipe is finally coated.

One way to protect against the corrosion of the pipe while it is
being shipped and stored above ground is to apply a "plant coating"
to the pipe as soon as possible after the pipe has been made. Sometimes the option to apply a plant coating is not available and
sometimes when available it is not the best choice to do so for both
technical and economic reasons.

Within the past three years research and development programs
carried out in conjunction with polyethylene tape coatings have
produced a liquid coating similar to the tapes primer for the protection of the surface of freshly shot blast cleaned pipe which will then
be exposed to the atmosphere for some time before the pipeline is
built. This storage coating is made from a butyl rubber compound
carried in organic solvent vehicle. It can be applied to the cleaned
pipe by spraying or with brushes and rollers.

The effectiveness of the storage primers has been tested and
proven in accelerated laboratory tests for a time equivalent to 3
years outdoor exposure with no significant change in the butyl rubber
based coating as applied to steel panels. Also, natural outdoor
weathering tests, still in progress, show no change in the primer
itself and complete protection of the pipe surface at the 18 month
time period.

The storage coating is completely compatible with polyethylene
tape and primers when they are applied before the pipe is finally
placed in the ditch. Although the storage coating may be completely
removed by shot blast cleaning before the tape is applied, it is not
necessary to do so. Cleaning prior to the application of a polyethylene tape is sufficient when any loose or spent storage coating is
removed. This means that standard over-the-ditch tape application
equipment and methods can be used when polyethylene tape is applied
over the pipe coated with the storage primer.

The technological importance of this storage coating for polyethylene tapes is that the desirable features of shot blast cleaning
which is done well in advance of the final coating operation, are
effectively preserved so that they can contribute to the in-service
performance of the polyethylene tape coating system.

Crosslinked Butyl Adhesives

The design and operating conditions of a large gas pipeline
planned for southern Europe are such that the engineers responsible
for selection of the coating wrote a specification for polyethylene
tapes requiring special adhesive bonding properties. The special
requirement is that the butyl adhesive used on polyethylene tape
must produce a minimum primed adhesion to steel pipe at 60°C of
2.0 Kg/5cm. The test method was specified as C.E.O.C.O.R. Peel
Adhesion Method B using a rate of peel equal to 2.5 cm per minute
(Ref. 7) The test specimen are sections of 50 mm diameter pipe
50 mm in length.

When it was determined that existing polyethylene tapes with butyl
adhesives could not achieve the specified level of primed adhesion at
60°C, a new adhesive was developed which did meet the specified value.

In order to meet this requirement a butyl adhesive system was
developed which retained a higher level of toughness and greater
resistance to shear and cohesive failures at 60°C. It was possible
to formulate and develop a butyl adhesive with these high temperature

qualities by adding chemical agents to the adhesive while it was being
manufactured which would later produce crosslinking within the butyl
rubber. The thrust of the R&D effort was to determine experimentally
the best chemical agent and the most effective amount of that agent.
Experimental trials were made in both pilot plant sized lots and
limited commercial volume.

Tables 2 and 3 - "Peel Adhesions at 60°C vs. Amount of Crosslinking Agent" are a compilation of the critical data for various experimental adhesives which were crosslinked with various amounts of one
chemical crosslinking system with their corresponding various levels
of peel adhesion to primed pipe at 60°C. These tables also show that
the desired properties of the crosslinked butyl adhesive are time/
temperature dependent. The desired adhesive properties are achieved
in a short time at 60°C and also over longer time periods at lower
ambient temperature. This means that pipe coated with polyethylene
tapes with the new adhesive at coating plants or over-the-ditch will
cure or crosslink in situ well in advance to the time the finished
pipe will reach the designated operating temperature.

This means that at the time of application the new butyl adhesive
has a desirable level of softness which enhances the qualities of
flow, conformability, initial bond level and interaction with the
pipes primer while later the toughness properties required at 60°C
are developed. This means that the adhesive is first moldable at the
time the tape is applied and then sets or firms up in place on the
pipe surface.

The final result is that the crosslinked adhesive on a polyethylene tape is physically stronger and more resistant to the changes
which can occur at higher operating temperatures. The crosslinked
adhesive has all of the desirable flow and bonding properties necessary for effective application at ambient temperature. Formerly it
was not possible to get such good high temperature performance and
low temperature application qualities from a pipeline coating tape
adhesive.

<u>The Correlation of Laboratory Test Results with Field Experience</u>

Several high molecular weight polymers, within the class of
modern plastics, possess the properties of high electrical resistance,
high resistance to moisture absorption and penetration, physical
toughness, and chemical inertness required in all materials which are
to be used as pipeline coatings. Polyethylene, a thermal plastic
polymer, possesses these critical and essential properties to a higher
degree than all of the thermal plastic and thermal setting materials
presently used as pipeline coatings. (<u>Ref. 8</u>)

The volume resistivity of polyethylene is 1×10^{16} ohm cm and
the moisture absorption is less than 0.01%. The polyvinyl chloride
and the epoxy resins used as pipeline coatings have volume resistivity
values no greater than 1×10^{11} ohm cm and moisture absorption as high
as .5 to 1.5%. (<u>Ref. 9</u>)

The thermal and chemical inertness of polyethylene is also
superior to those of other materials used in pipeline coatings.

There are many laboratory tests for pipeline coatings which are
used to give an indication of how materials may perform on underground pipelines. Of all of the pipeline coating laboratory tests,
it is most likely that tests performed on coated pipe specimen to
evaluate the electrical resistance properties while the coating is
exposed in aqueous solutions may best correlate with inservice performance.

The performance of polyethylene pipeline tape coatings in laboratory and field tests and in-service on actual pipelines shows that there is a correlation between electrical resistivity type laboratory tests and in-service coating resistance.

For example, a polyethylene tape coating was wrapped on pipe specimen which were tested at the Engler-Bunte Institute at the University of Karlsruhe by the DIN 51558-59 test method for coating resistivity. The test results after one day and 100 days continuous immersion in three electrolytes consisting of three different inorganic, aqueous solutions and the test performance by the DIN 51558-59 test method are given in Table 4.

As another example, Gaz de France, the major gas pipeline operators in France, tests the electrical resistivity of pipeline coatings in the laboratory by immersing a U-bend shaped coated pipe specimen in a 15% NaCl solution for a period of 200 days and measuring the coating resistance at periodic intervals. (Ref. 10) The exposed area is kept constant so the test results are expressed simply in ohms. The resistance of a polyethylene tape coating on the U-bend pipe specimen after 200 days in this Gaz de France test is 5×10^{11} ohms.

These two laboratory tests prove that polyethylene-based tape coatings have very good electrical resistance qualities initially and after prolonged testing which are consistent with the known dielectric properties of polyethylene.

While laboratory conditions represent conditions of ideal application and controlled environment, a pipeline coating applied under commercial conditions and exposed in natural soils encounters many variables of application and of soil properties which act in concert to detract from the inherent properties of the coating itself. This is taken into account by coating and cathodic protection engineers when they rate the electrical resistance qualities of a pipeline coating on operating pipelines. As an industry rule of thumb, the performance of a coating is rated as excellent if its electrical resistance as measured on an underground pipeline after the line is in service falls within the range of 50000 to 5000 ohms per square meter. With a stable pipeline coating, the measured coating resistance will remain constant with the passage of time.

In one test for coating resistance, a section of 610mm diameter pipe 37.2 km long was coated with a polyethylene pipeline tape coating. This section of pipe is used as a field test designed for measuring coating resistance so that the 37 km of 610 mm pipe can be tested yearly (Ref. 11). After 16 years, the coating resistance of the polyethylene tape has not changed from its original value of 31,000 ohms M^2.

On another operating pipeline coated with a polyethylene tape and consisting of 440 km of 300 mm and 400 mm pipe, coating resistance measurements were made as part of the engineering survey for cathodic protection. The United States based consulting and engineering firm who conducted the coating resistance tests of the polyethylene tape for the entire 440 km reported that the coating resistance was greater than 30,000 ohm/M^2 with several sections of ohms per square meter. (Ref. 12)

The engineering company's report to the pipeline operating company stated the following about these high coating resistance values:

"The (Coating Resistance) values are particularly exceptional considering the pipeline has been installed

and operating about one year before most of the test
data were obtained. During that period, the area had
received above average rainfall, the pipe had been loaded
with product, and subjected to cycles of pressure,
temperature, velocity, soil moisture variations and other
factors which tend to promote damage to coatings."

The two cases of in-service coating resistance for pipelines
coated with polyethylene pipeline tape coatings and the two laboratory
test results cited previously reflect the known high electrical resistance properties of polyethylene.

Since water was always present in the laboratory tests and in the
soil surrounding the tape coated operating pipelines in the tests for
coating resistance reported above, a direct correlation can be drawn
between the laboratory and field results obtained. In addition, it
can be stated that the high resistance of polyethylene to moisture
absorption and moisture penetration are also imparted to and maintained
in pipeline tape coatings. If moisture had been absorbed or had
penetrated the polyethylene tapes in the laboratory or in the field,
the coating resistance values would have been much lower.

"ARCTIC-WEATHER" APPLICATION OF POLYETHYLENE TAPE COATINGS

The discovery and then the production of crude oil and natural
gas in the Arctic and sub-Arctic regions of the world has led to the
planning and construction of the pipelines necessary to bring the oil
and gas to markets to the south. According to a report in the Construction Progress section of the October 1971 issue of Pipeline Industry entitled "Special Equipment, Methods for Far-North Construction"
the building of the northern portions of these pipelines will require
materials, equipment and construction methods which take into account
the extremely cold weather conditions likely to be encountered.
(Ref. 13) The staff writer for Pipeline Industry covers in this report
many of the phases involved in the building of pipelines in the Alaska/
Canada Far-North. The problems encountered during cold weather construction are discussed from ditching to welding, to coating and lower-
ing-in.

Pipeline coating materials and application techniques used in cold
climates, Pipeline Industry points out, must have properties which are
compatible with the conditions brought about by long exposure to
temperatures in the -29° to -45°C range. When pipeline coating
materials are exposed to low temperatures for prolonged periods of
time, the predominent effect on their physical properties is a tendency
to decrease their flexibility and pliability resulting in brittleness.
This is to say that the plastic nature of a coating material, if it be
coal tar enamel, epoxy, or polyethylene can change so that the material
becomes more glass-like as its temperature decreases.

Each of the commonly used pipeline coating materials is affected
to a different degree by lowering temperatures. Some are too stiff
and brittle even at the freezing point of water (0°C) while others retain a workable degree of flexibility and pliability at temperatures
below -40°C. Other factors being equal, it then follows that the
coating material least affected by low temperatures may be the one most
suitable for use on pipelines built in the Far-North. Table 5 shows
the relative brittleness temperatures of coal tar, epoxy, polyvinyl
chloride and polyethylene.

Pipelines intended for construction under arctic-winter conditions
can be coated at coating plants or over-the-ditch. In either case,
the coating material must maintain a practical degree of flexibility

and pliability at sub-zero temperatures so that it will resist damage
which can occur as the coated pipe is flexed, bent and bumped during
handling.

Flexibility and pliability at low temperatures are more important
for plant coatings used in sub-zero weather because the pipe with the
coating already on it is subjected to the rigors of the bending operation along the right-of-way, while the coated pipe's temperature is
equal to the prevailing ambient temperature. This is in addition to
the other extra handling steps required to get a pipeline using plant
coated pipe into the ditch and backfilled. When coating the pipe
over-the-ditch, the bare pipe is bent before it is coated and the pipe
after it is coated is handled the minimum amount before backfilling
takes place.

Based on field experience and laboratory testing polyethylene
tape type pipeline coatings appear to have the properties for application at plants or over-the-ditch wherever construction conditions
include very cold winter weather. The previously cited report in the
October 1971 issue of Pipeline Industry as well as an article by
D. Deasen entitled "Large Diameter Projects Offer Experience for
Arctic Pipelines" published in the August 1972 issue of Pipeline
Industry, refer to some successful experience with polyethylene tape
coatings on pipelines built during sub-zero weather. (Ref. 14)

Polyethylene tape pipeline coating show suitable performance
properties over a range of negative temperatures in laboratory tests
and actual field experience.

Laboratory Tests

Some laboratory tests conducted at low temperatures are used to
develop data about the performance characteristics of pipeline coatings
for application to pipelines built during severe winter weather. The
ability of coating materials to remain flexible and pliable at low
temperatures can be judged by consideration of the results obtained
from impact and bending tests.

The impact resistance of a polyethylene tape system, a polyvinyl
chloride tape system and fusion bonded epoxy system were determined
by performing a drop weight impact resistance test on these coatings
while the test sample and apparatus were maintained and operated at
temperatures in the range of 21°C to -40°C. The test method used was
based on the ASTM G-14 single point impact resistance test method.
The results of these tests are given in Table 6.

The test results in Table 6 show two important facts about the
polyethylene tape coating. These are:

    1)   The impact resistance is greater at temperatures
        below 21°C.

    2)   When failure occurs, the point of damage is confined
        to the area directly under the impacting object and
        the polyethylene showed no fracture lines away from
        the impact point.

The polyvinyl chloride tape and the epoxy fusion bonded coating test
results show that fracturing took place away from the point of impact
thereby increasing the size of the holiday. The fracturing indicates
that the PVC and epoxy are brittle at all test temperatures below 21°C.
The fact that the polyethylene tape did not fracture means that this
material did not exhibit brittleness at any temperature including
-40°C, and that the coating damage was confined to the area of the

the exact impact point.

Bending tests are often used to determine the flexibility and brittleness of pipeline coatings. When the bending tests are performed over a range of temperatures from normal ambient to sub-zero, the effect of depressed temperatures on these properties is evaluated. A bending test was performed on a polyethylene tape pipeline coating at 23°C and -40°C to see how this coating material was affected by low temperatures.

The test consisted of applying strips of the polyethylene tape to primed thin flat steel panels. The dimensions of the steel panels were 6.25 cm x 30 cm x 0.42 cm. The thin gauge coated steel panels were used to simulate actual coated pipe because it is easy to bend the flat panels to any degree desired. The panels with the tape on one side were conditioned and bent through a 90° angle while the test panels were held at 23°C and then -40°C. The polyethylene tape which was bonded to the outside of the 90° bend did not crack or craze but remained totally flexible at 23°C and at -40°C. The degree of the bend in this test far exceeds the flexual bending or permanent bending a coated pipe will undergo during construction.

The results of the laboratory impact and bending tests indicate that polyethylene tapes remain plastic at very low temperatures and because of this they are suitable for use under Arctic pipeline conditions.

The good low temperature performance of polyethylene tape coatings as shown by laboratory tests and also actual field application in arctic-like winter weather is attributable to the inherent low brittleness temperature of polyethylenes. The Encyclopedia of Polymer Science and Technology reports that the brittleness temperature of the types of polyethylene resins used in polyethylene pipeline tapes fall in the range of -55°C to -80°C (Ref. 15).

Field Experience

During the years of 1965 to 1974, polyethylene tape pipeline coatings were applied over-the-ditch to 2832 kilometers of pipelines built during Arctic-like weather conditions. Twenty-two separate pipelines were involved covering a pipe diameter range of 50 mm through 900 mm. 950 km are 900 mm, 128 km are 750 mm, 419 km are 600 mm and 376 km are 300 mm pipe. Table 7 gives the dimensions of these pipelines and the winter in which they were built. (Ref. 16)

The temperatures during construction of these pipelines was very often below -29°C. In one case in the Canadian Northwest Territory, the polyethylene tape pipeline coating was applied while the air temperature was -47°C.

The application and construction techniques used on these lines were modified in two ways from the normal summer time methods. The two modifications are that the tape was kept warm in a heated sled at about $+5^{\circ}$C up to the point of application. The other was that gas fired line travel heaters were used when needed to dry the frost and ice out of the pipe ahead of the tape wrapping machine. Average coating rates of 2.3 to 3.2 kilometers per day were experienced for these pipelines. All of the normal pipe laying and coating operations were performed without damage to the polyethylene tape attributed to the cold temperatures.

The performance of the polyethylene tape coatings after burial on these lines is reported to be very satisfactory. Current require-

ment data on two of the lines is reported in Table 8. The current
requirements in each case is very low (17.1 and 24.6 microamps/sq.
meter)after 10 and 16 years in service. These current requirements
indicate the coatings were in very good physical condition when the
lines were built and have remained so during the year to year operation
of the pipelines.

Polyethylene tape pipeline coatings have been applied at coating
plants to pipe used during sub-zero construction. A 75 kilometer
600 mm pipeline using a 1.34 mm thick two layer polyethylene tape and
outerwrap plant coating system was built in the provinces of Ontario
and Quebec, Canada during the winter of 1976-77. The coated pipe was
trucked, strung, bent, welded and placed in the ditch when air temperatures were in the -20°C to -30°C range. The polyethylene tape
plant coating was not damaged when the coated pipe was bent at these
low temperatures. The plant coating remained smoothly bonded to the
pipe without the occurence of any brittle fractures in the polyethylene
tape and outerwrap throughout the entire winter construction period.

Laboratory and field experience with polyethylene tape pipeline
coatings at temperatures considerably below normal ambient conditions
show that this type of pipeline coating can be used successfully on
pipelines built under Arctic and sub-Arctic winter weather conditions.
They can be applied successfully over-the-ditch or at coating plants
and will resist damage due to the fact that the polyethylene does not
become brittle at temperatures as low as -40°C. The current requirements for polyethylene tape coatings on pipelines built in Arctic type
winter weather are proving to be very low after many years in-service.

<u>References</u>

1. O'Donnell, J.P. : "Coal Tar Enamel Still Most Widely Used
    Pipeline Coating in Industry". The Oil and Gas Journal. (January,
    1977).

2. Ferry, John D. "Viscoelastic Properties of Polymers". John
    Wiley & Sons, Inc. (1970)

3. Fifth Symposium on Line Pipe Research, Pipeline Research.
    Committee of AGA, Houston, Texas, November 20-22 1974, American
    Gas Association, Arlington, Virginia, Catalogue No. L30174

4. Fessler, R.R. "Combinations of Conditions Causes Stress-Corrosion
    Cracking", The Oil and Gas Journal, 81-83 February 16, 1976.

5. Payer, J. H., Berry, W. E. and Boyd, W.K.: "Constant Strain Rate
    Technique for Assessing Stress-Corrosion Susceptibility", ASTM
    STP-610, American Society for Testing and Materials, Philadelphia,
    Pennsylvania, 1976.

6. Berry, W.E., Barlo, T.J., Payer, J.H. and Fessler, R.R.:
    Corrosion/'78 Paper No. 64. National Association of Corrosion
    Engineers (NACE) Conference 1978, Houston, Texas.

7. C.E.O.C.O.R Commission 2, "Draft Recommendation for External
    Coating by Means of Polyethylene Applied by Powder Fusion or
    By Extrusion on Steel Tubing and Fittings for Underground Pipelines"
    C2/8/, June 1978.

8. Harris, G. M.: "The Corrosion Protection of Underground Pipelines
    by Plastic Tape Pipeline Coatings" Paper A 4 Proceedings of The
    Second International Conference on the Internal and External
    Protection of Pipes" Organized and sponsored by BHRA Fluid
    Engineering, Cranfield. Held at University of Kent at Canterbury
    England, Spetember 7-9, 1977.

9.   <u>Modern Plastics Encyclopedia</u>, <u>1967</u>

10.   <u>Christian, R.</u>: "Protection of Buried Pipe Connections by Use of Tapes" (Protection externe des canalisations onterrées par bandes adhesíves), Assoceation Technique de L'endustrie Du Gaz En France" Congrés 1963.

11.   <u>Seager, W. H.</u> "A Review of Cathodic Protection Current Requirements Interprovincial - Lakehead Pipeline System May, 1978" Commonwealth Seager Group, (unpublished).

12.   "Canyon Reef Carriers, Incorporated Sacroc Carbon Dioxide Pipeline Cathodic Protection Completion Survey" Ebasco Services Incorporated, Houston, Texas, August, 1973 (unpublished).

13.   "Special Equipment for Far-North Construction" Pipeline Industry, (October, 1971).

14.   "Large Diameter Projects Offer Experience for Arctic Pipelines", Pipeline Industry, August 1972.

15.   "Encyclopedia of Polymer Science and Technology" John Wiley and Sons, Inc. (1971).

16.   <u>Harris, G. M.</u> "The Performance of Plastic Tape Pipeline Coatings at Sub-zero Temperatures", Corrosion/'79 Paper 213, National Association of Corrosion Engineers (NACE) Conference 1979 Atlanta, Georgia March 12-16, 1979.

TABLE 1.

Test results obtained from testing SCC inhibition derived from Polyethylene tape primers containing sodium tripolyphosphate. Test
Method - Battelle Institute slow strain rate stress method (using X-52
pipe steel)

Specimen No.	Time to Failure (Hrs)	Reduction in Area (%)
1	23.05	50.3
2	24.12	52.1
3	21.37	55.0
4	21.62	51.7

Note: The X-52 steel without the presence of an inhibitor fails
in less than 20 hrs. with a 40% or less reduction area.

TABLE 2.

Peel Adhesion[a] at 60°C vs. Concentration of Crosslinking Agent

Aging[b] Time (Hr)	Peel Adhesion (kg/5 cm) Conc. of Crosslinking Chemical		
	0%	0.8%	1.4%
5	0.95	1.68	1.68
17	1.34	2.11	2.54
66	1.66	2.99	2.95
336	2.04	3.50	4.49
672	1.77	3.47	4.77

(a) Peel rate at 2.54 c/min; peel angle 90°; test temp. at 60°C.

(b) Aging condition at 60°C.

TABLE 3.

Peel Adhesion[a] at 60°C vs. Concentration of Crosslinking Chemical

Aging[b] Time (Hr)	Peel Adhesion (kg/5 cm) Conc. of Crosslinking Chemical		
	0%	0.8%	1.4%
0	0.95	1.68	1.68
336	1.75	2.32	2.79
672	1.48	2.77	2.77

(a) Peel rate at 2.54 cm/min; peel angle 90°; test temperature
at 60°C.

(b) Aging condition at 22°C plus 5 hours at 60°C.

TABLE 4.

Resistivity of Polyethylene Tape for DIN 51558-59

Solution	Resistivity (ohms/M^2)	
	1 Day	100 Days
.1 N HCl	9.2×10^{10}	1×10^{11}
.1 N NaCl	1×10^{11}	7×10^{10}
.1 N NaOH	1×10^{11}	1×10^{11}

TABLE 5.

Brittle Temperature of Four Pipeline Coating Materials

Coating	Brittle Temperature
Coal Tar Enamel	$0°C$ to $-20°C$
Polyvinyl Chloride Tape	$0°C$ to $-20°C$
Epoxy, Fusion Bonding	$0°C$ to $-20°C$
Polyethylene Tape	Below $-40°C$

TABLE 6.

Impact Resistance[*] vs. Temperature (inch-pounds)

Temperature ($°C$)	Polyethylene Tape System (2.25 mm)	P.V.C. Tape (1.25 mm)	Fusion Bonded Epoxy (.55 mm)
21	72	26	18
-2	102	40 f	64 f
-18	92	50 f	72 f
-29	102	15 f	64 f
-40	90	16 f	64 f

f = Holiday and fracture of coating away from point of impact.

* A falling weight single point method similar to ASTM G-14.

TABLE 7.

Extreme Cold Weather Construction Experience with Polyethylene Tape
Pipeline Coatings

Pipeline No.	Pipeline Size km	Dia (mm)	Year
1	128	750	1965
2	63	300	1967-69
3	304	400-500	1968-69
4	112	300	1968-69
5	40	200	1968-69
6	29	450	1969-70
7	40	150-200	1969-70
8	58	800	1969-70
9	88	800	1969-70
10	48	300	1971
11	16	250	1972
12	48	150-250	1968
13	26	150	1972
14	90	400	1969
15	64	200	1969
16	112	200	1969-70
17	240	600	1967
18	293	300	1969-70
19	390	800	1971-72
20	300	800	1973
21	102	800	1973-74
22	53	300	1969
23	179	600	1971

2832 km

TABLE 8.

The Performance of Polyethylene Tape Pipeline Coatings Which Were
Applied During Extremely Cold Weather (Commonly Below -29°C)

(Current Requirements)

Pipeline Number From Table 2	Construction Date	Last Survey Date	Microamps per sq.Meter
#17 (240km/600 mm)	1967	1976	17.1
#18 (293km/300 mm)	1969-70	1976	24,6

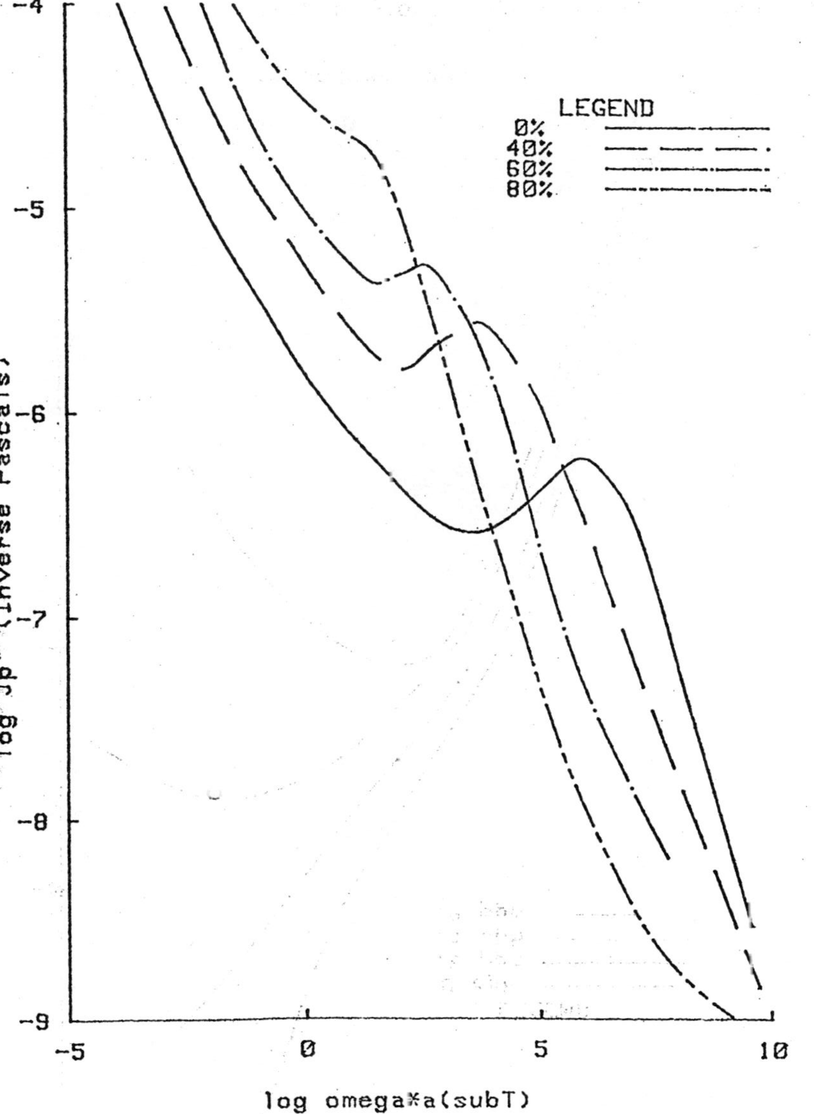

Fig. 1   Changes in loss compliance J" of an adhesive
as the amount of tackifier increases from 0% to 80%.

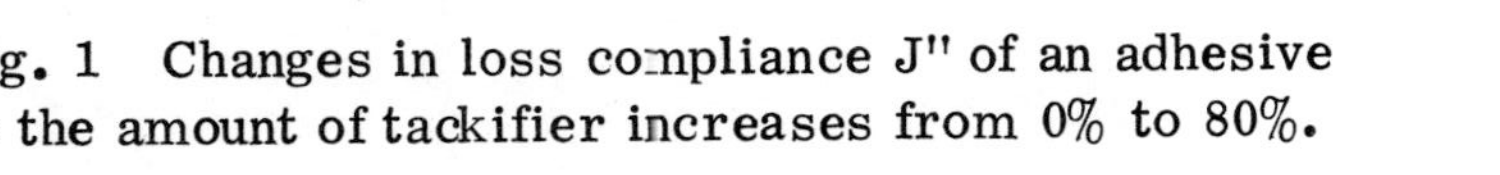

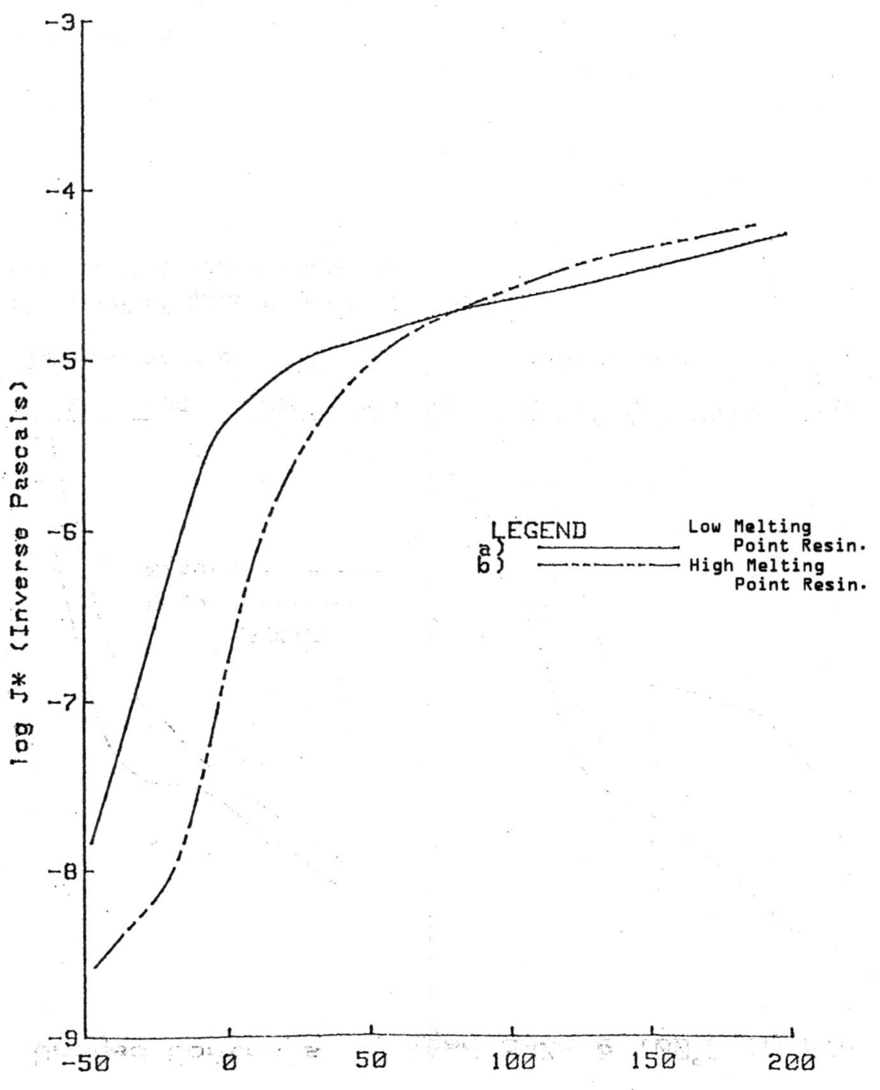

Fig. 2   Changes in glass transition (Tg) due
to tackifiers with different melting points.

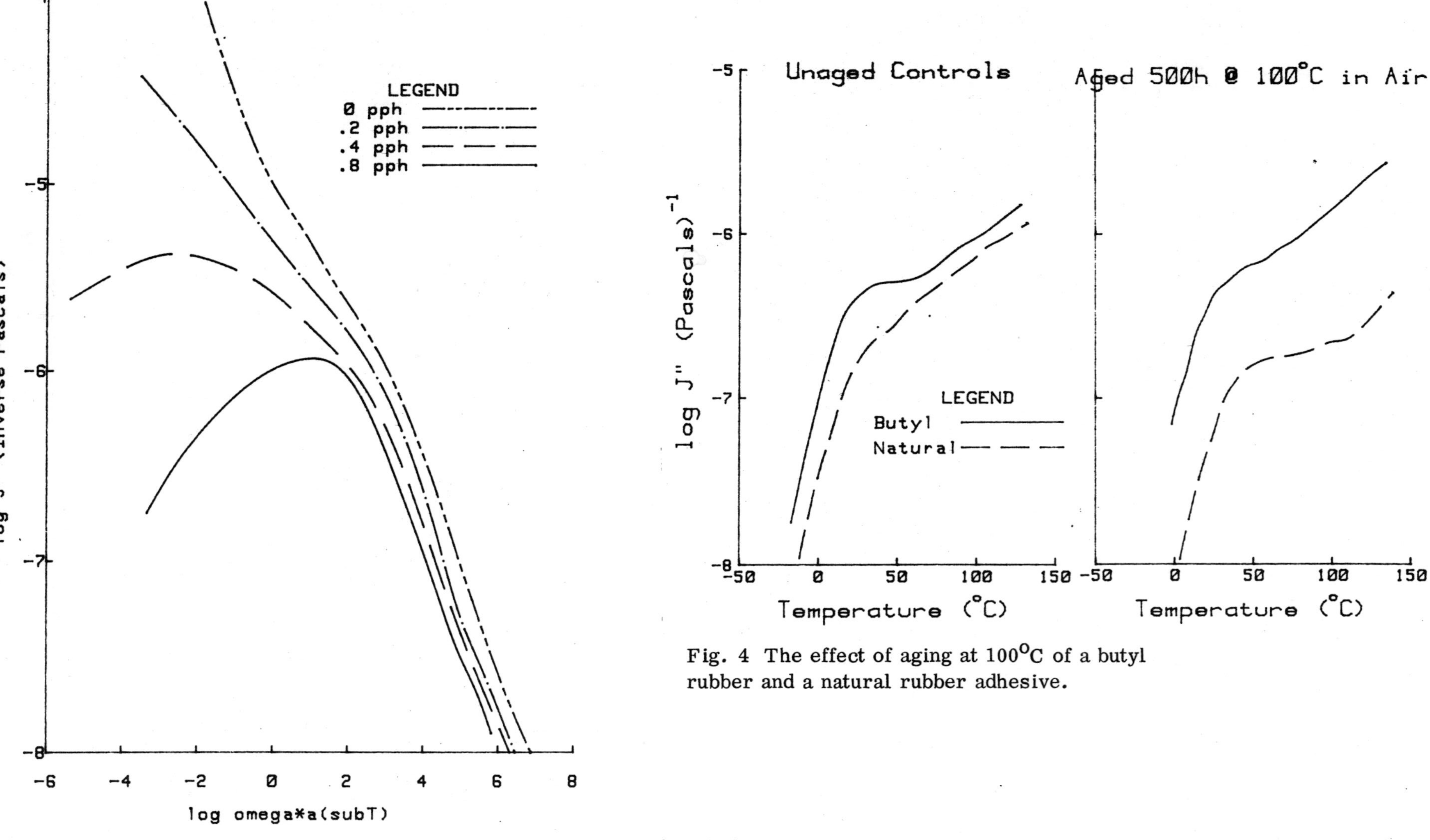

Fig. 4 The effect of aging at 100°C of a butyl rubber and a natural rubber adhesive.

Fig. 3 Effects of crosslinking on an adhesive system containing an elastomer (Me 80, 000) and a chemical crosslinking agent in amounts varying from 0 pph to 0.8 pph.

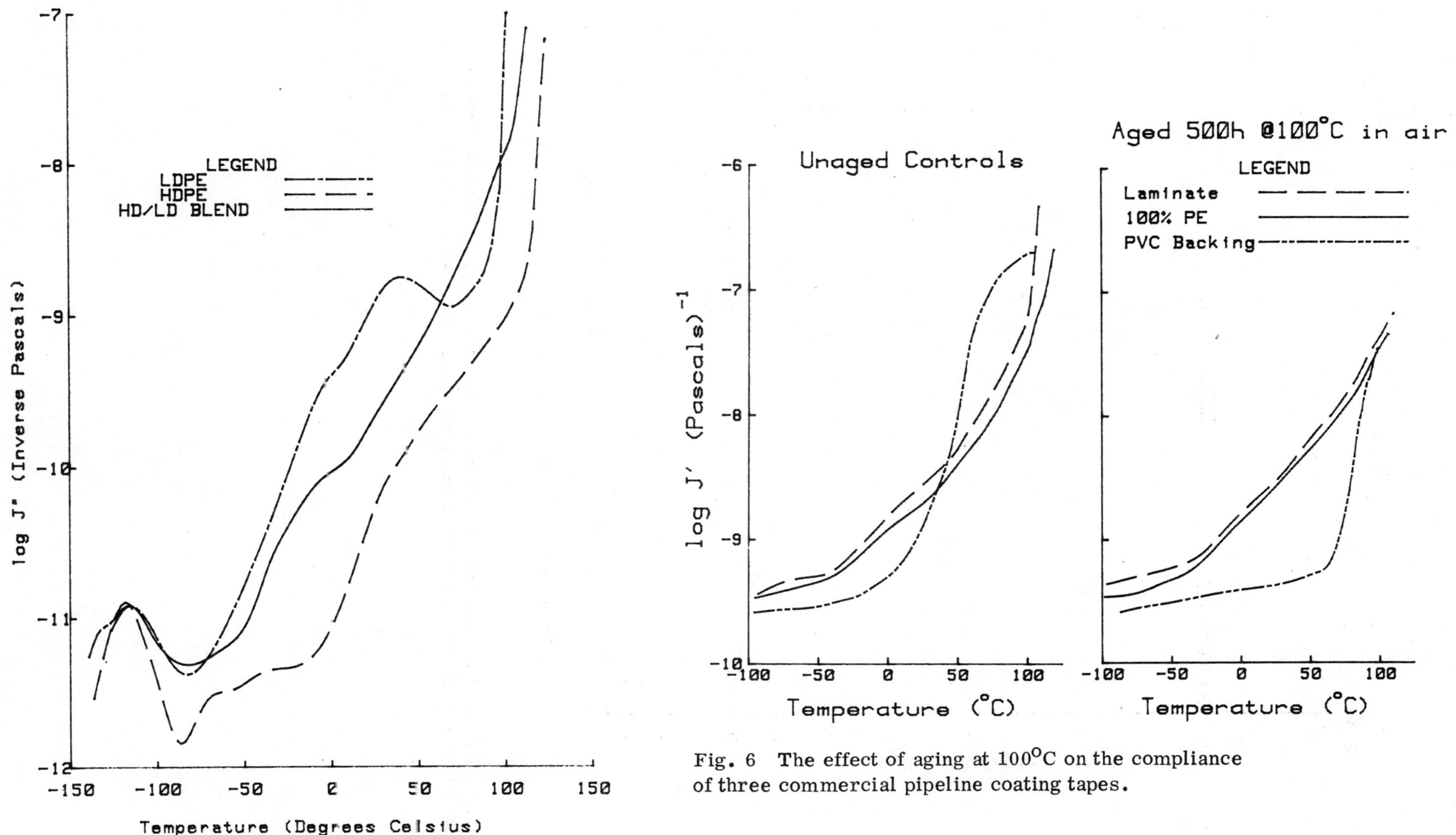

Fig. 6   The effect of aging at 100°C on the compliance of three commercial pipeline coating tapes.

Fig. 5   The compliance of three polyethylene tape backings containing (1) 100% low density polyethylene, (2) 100% high density polyethylene and (3) a blend of low and high density polyethylene.

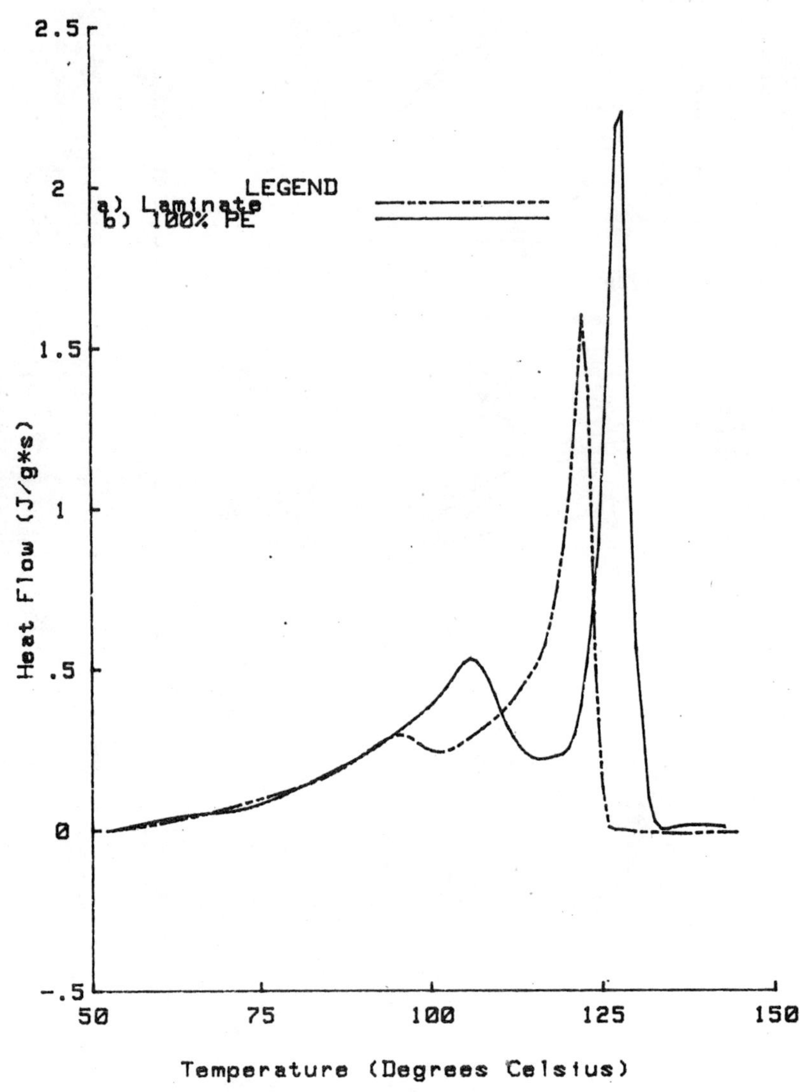

Fig. 7   Differential scanning calorimetry curves for two polyethylene tapes - one is pure polyethylene, the other has a non-polyethylene anchoring layer.

CEMENTITIOUS TAPE – A NEW CONCEPT IN EXTERNAL PROTECTION OF PIPES AND PIPE INSULANTS

R.S. Whitehouse, PhD, CChem, MRIC, FRMS, and G.E. Richardson, HNC Chem.

Evode Limited, U.K.

Summary

Wherever pipework is used, the need for its protection or insulation is a major consideration. There have been many developments in the materials used in these aspects of pipe protection, together with the development of sophisticated quality control measuring devices to ensure that the pipe is adequately protected. Rapid advancement in polymer technology has seen the introduction of such systems as pressure sensitive tapes, shrink wrapped sleeves and fused powder. The body of this paper is concerned with the use of Portland cement, compounded into tape form, as a protective medium for steel pipes and also as a protection for pipe insulation. The development of this tape, its testing and applications, are discussed and comparisons drawn against some established protective systems.

Held at Imperial College, London, England.
Organised and Sponsored by BHRA Fluid Engineering

<u>Introduction and Background</u>

Ever since pipes have been used to transfer material from one location to another there has been a need to protect the pipe composition from corrosion, particularly external oxidative corrosion, or to protect pipe insulation from environment and accidental mechanical damage. With the rapid advances in polymer technology the types of external protection have increased. Such systems as coal tar pitch, pressure sensitive tapes, shrink wrapped sleeves, fused powders and paints are now extensively used. They all depend however, on providing a continuous liquid and/or vapour barrier between the environment and the clean pipe surface or insulating material. The problem of providing a continuous impervious coating is obviously very difficult in practice and this has led to the development of sophisticated quality control measuring devices. In recent years cathodic protection has been used extensively to provide a duplex protective system with the above systems.

We feel the most difficult problems associated with protective coatings are :

1. Providing the impervious coating.

2. Application.

3. In-situ repair.

A further consideration in each of the above areas is one of cost of materials and labour. It may often prove that where a relatively cheap material is being used, the manpower required to apply that material, often by specialist contractors, results in a system which is low in materials cost, but very expensive when applied.

Instead of developing another sophisticated polymer system with equally sophisticated and critical application characteristics, we decided to consider a material which has been used successfully for many years in the construction industry. That material is Portland cement. Cement provides an alkaline environment adjacent to the pipe surface, thus inhibiting normal ferrous corrosion. It does not have to provide an impervious barrier, in fact the opposite is beneficial to minimise pressure/ temperature gradients by conventional diffusion. Furthermore it sets to a hard, impact resistant layer which is another important feature when selecting pipe protective systems, especially when such systems are used in harsh conditions where failure of the system may be directly accountable to impact damage. Resistance to impact or mechanical damage is of major importance especially when pipe insulating materials such as polystyrene, phenolic, isocyanurate and polyurethane foams are used, these by their very nature being very prone to this type of damage with resultant loss of insulation performance. Cement alone can only be used for above ambient temperature insulation. Below this temperature a vapour barrier coating is also required to stop ingress of moisture and ice formation in the insulation.

Cement or mortar has certain deficiencies such as poor flexural strength, low adhesion to metal and difficult application properties. The use of cement or mortar involves the use of mixing equipment, there are handling problems in transporting the material to the place of work, it often needs keying to the pipe surface to optimise adhesion and it requires skilled operatives to either spray or trowel the material into place in order to obtain an accurate degree of cover with an acceptable finished appearance.

By combining our cement and coatings technology we have developed a cement impregnated tape. This provides a cementitious layer which is reinforced by a special fabric membrane and a mixture of polymer additives which combine to give excellent adhesion, compression and tensile strength and resilience. The product is supplied in roll form in a range of widths to suit the particular job in question. Due to the material being fully compounded during its manufacture, no on-site mixing is required; the material can be taken to the actual point of application where it only requires a 20 - 30 seconds immersion in water to activate the cement and render the tape supple for direct application to the pipe or insulant surface. Once applied the tape has excellent wet adhesion to allow efficient placing around the pipe surface, both on straight sections and complex bends. The tape will then cure to provide a hard but resilient film which will accommodate small amounts of movement and pipe vibration, without loss of adhesion or cracking.

The construction of the tape consists of a woven cotton membrane which acts as a support and carrier for the cement. Being cotton it possesses excellent wet strength and in the cured product state is resistant to biological attack. The weave of the membrane is such that it not only acts as a support but also as a reinforcement for the cement when cured. The membrane has two directional properties which offer considerable reinforcement to the cured tape; it is significantly more efficient than the random fibres in FRC. In the production of the finished tape, the membrane is both impregnated and coated by the compounded cement and polymeric binders which are held in a non-aqueous dispersion. The final thickness of the tape is 1 mm. The blend of polymeric binders with cement allows the non-aqueous coating of the cement so that no hydration of the cement takes place until it is activated in the water; thus maximum strength and performance are obtained from the cement.

There are two methods of application, which depend on the size of the pipe being considered.

Firstly, for small diameter pipes of short length or for wrapping around bends etc., the SPIRAL WRAPPING technique is employed. Roll widths of 25 - 150 mm are ideally suited to this method. The appropriate length of tape is torn from the roll and the polyethylene interleaf placed to one side for later use. The tape is then activated in water for 20 - 30 seconds during which time sufficient water is absorbed by the tape to activate the cement. The tape is then removed from the water and the excess water allowed to drain, after which the tape is applied at approximately 45° to the pipe axis and wrapped spirally around the pipe, allowing 25 mm overlap to the previous layer. Consolidation of the surface is achieved by smoothing over with gloved hands.

For larger diameter pipes of large runs the CIGARETTE WRAPPING technique is more appropriate, tape widths up to 1 metre in length are suitable for such applications. In this method a length of tape equal to the circumference of the pipe, plus 50 mm for joining, is removed from the roll. Activation is as previously described and the activated tape applied to the pipe circumferentially. Successive sheets are then applied in a similar manner allowing a 50 mm overlap join to the previously applied material.

In both methods of application a uniform covering to the pipe is simply and efficiently obtained without the need for skilled labour or special equipment. Due to the fact that in each case overlap joins are used, total coverage is afforded to the pipe. Should these joins be imperfect, the fault can be identified easily by visual inspection. Once cured a continuous coating is formed with no interfacial weakness between overlapped layers of tape.

Cement cures by chemical hydration reactions and to allow these reactions to go to completion, a certain amount of water has to remain in contact with the cement for up to 48 hours. This may be considered as the only drawback to this particular system in that by applying thin layers of cement the water may dry faster by evaporation than the time required to hydrate the cement. There is no problem if pipes are to be buried below ground providing there will be sufficient water in the soil to allow the curing reaction to take place. During warm conditions, however, especially in boiler rooms, pipe ducts or tropical climates, this loss of water by evaporation will be too fast to allow full hydration of the cement. This is where the discarded polyethylene interleaf from the roll comes into use. It will be remembered that when the required amount of tape was taken from the roll, an equivalent length of polyethylene interleaf was also removed. This is then used to cover the freshly applied tape and is left in place for 36 - 48 hours so that sufficient moisture is retained to allow complete hydration of the cement. After this period of time the polyethylene is simply pulled off from the surface of the hardened cured product. It may be left in position in, for example, underground applications where this will not be noticed or be of any detriment to the wrapping.

Typical properties of the cured product are as follows :

Tensile strength   2.6 MN/m^2 parallel to membrane
(multiple layer specimens)   =

Compressive strength   =   26.3 MN/m^2 parallel to membrane
(multiple layer specimens)   15.6 MN/m^2 perpendicular to membrane

Vapour permeability   =   195 g/m^2/day - single thickness at 75% rh
(BS 3177 : 1959)   498 g/m^2/day - single thickness at 98% rh

Fire resistance   =   Indicated Class 1 rating to BS 476 :
                                            Part 7 : 1971

                                        Indicated Class 0 rating to BS 476 :
                                            Part 6 : 1968

Chemical resistance   =   Unaffected by aliphatic hydrocarbons,
                                          common industrial solvents, vegetable
                                          and mineral oils, and diesel fuel.

                                        It is seriously affected by mineral acids.

<u>Laboratory Testing</u>

1. <u>Compression - Tension Tests</u>

    These tests were performed according to ASTM specifications C 109-72 (Tensile strength of Hydraulic Mortars) and C 109-73 (Compressive strength of Hydraulic Mortars). A comparison was made between a standard mortar and the cement tape. Specimens were cured for 0, 3, 7 and 28 days in a wet box before testing at 28 days using a Tinius Olsen machine at a test rate of 3 mm/mm/min. (Fig. 1.1).

    Results show that the optimum curing procedure was 3 days in the wet box, followed by drying at ambient temperature and humidity.

    Standardising on this procedure, further results were obtained as shown in Fig. 1.2 (load perpendicular to cotton). Forces applied parallel to the textile fibres gave an approximately 35% decrease in strength.

    Further evaluation of these results led to the following conclusions :

    (a) In compression the product does not separate or crumble at failure but retains its general shape. This is unlike mortar which crumbles at failure.

    (b) The reinforcing membrane in the product allows it to take greater strains in compression and tension. This results in lower modulus of elasticity when compared with concrete.

    (c) On a strength for strength basis the tape product is 35% lighter than mortar per unit volume.

2. <u>Impact Resistance</u>

For pipe wrapping applications, if the material is to provide satisfactory
performance, some degree of impact resistance is required. Impact resistance
is the ability of a material to absorb energy without failure occurring. In
these tests the Charpy 'V' notch test was used to determine impact resistance.
This test method was originally developed to determine the energy absorbing
characteristics of metals and later adapted to test plastics. Impact tests
are not normally performed on concrete specimens as the impact resistance of
concrete or mortar is almost non-existent. The reasons for this are that
concrete or mortar, unreinforced or in thin sections, are weak in tension
and have many flaws which cause failure due to crack propagation. The flaws
act as stress concentration and so promote crack propagation. The wide range
of particle sizes also leads to stress concentration and low impact resistance.

The reasons for the tape product having a degree of impact resistance are that
the fibres in the supporting membrane reinforce the cement, acting as a stress
distribution and damping medium.

In the tests samples 70 mm long × 10 mm × 10 mm cross section were produced
and cured as before for the tension/compression tests. The impact resistance
before failure was measured by striking the samples with a 27.3 kg swinging
hammer producing 150 J of available energy. The apparatus used was as shown
in Fig. 1.3.

Load versus time measurements were taken using an oscilloscope, a typical trace
being obtained as in Fig. 1.4.

The observations made from the results of these tests are as follows :

(a) Cracks although starting quickly propagate only very slowly.

(b) Impact failure is not truly ductile as shown by the fairly sharp peaks
     on the oscilloscope trace.

(c) Impact resistance is dependent on impact speed and in the tests performed
     an impact resistance of 115.0 kg to 140 kg was obtained.

(d) The maximum point of the oscilloscope trace is not very sharp depicting
     a slow fracture.

3. <u>Corrosion Resistance</u>

The method used was that as defined in ASTM C464-64 for corrosion of steel and
involved the following procedure :

6.25 mm × 50 mm × 50 mm cold rolled steel was buffed to produce a smooth,
bright and shiny surface. All oil and surface contaminants were removed with
mild abrasives and detergents. Sufficient similar sized pieces of tape were
cut, activated in water, and built up on each of the steel specimens to
produce a layer approximately 62.5 mm thick on the steel. Specimens were then
air dried for a period of 14 days as specified. After this period of time the
cement product was removed from the test surface and the following results
noted :

(a) corrosion in the form of rust was found on the external edges of the steel
     but according to the ASTM specifications this is ignored as edge effects;

(b) over the whole of the test area there was only a minimal amount of rust
     up to 15.6 mm from the edge boundary with minimal scaling and comparable
     results were obtained with the ASTM specified cement. This degree of
     corrosion resistance with the tape product is not only attributable to

the alkalinity of the cement but also to the tighter bond of the tape
product to the test surface as compared with the cement.

4. Tests with Cathodic Protection

Cathodic protection of pipes is sometimes used as a secondary protection
to pipe wrapping materials. One of the problems that may arise with this
method is that disbondment of the pipe wrapping can occur when used in conjunction with cathodic protection methods. In order to assess if the tape
product could be used in this application the following procedure was carried
out.

Cement tape was applied to a steel pipe abraded to SA $2\frac{1}{2}$ (Swedish Standards
Institution 515 05 5900 – 1967) and allowed to cure. A hole was then drilled
through the cement product (3 mm in diameter) and the pipe immersed in a 3%
sodium chloride solution which was changed every 3 days during the test. A
potential of 1.5 – 1.8 V was applied across the electrodes (see diagram) and
the test continued for a period of 21 days. After this period of time the
cement-tape covered pipe (cathode) was examined for any signs of disbondment,
but none could be observed. Tests terminated after 110 days when no signs of
disbondment were observed (Fig. 1.5).

Discussion of Applications

In the field of pipe work the cement tape product appears to offer certain useful
features.

(1) Steel Pipes

Due to the alkalinity of Portland Cement and the dense, tightly bonded surface
of the tape product, the material may be used as a corrosion protection for
steel pipes. In below-ground applications, the coating will be more resistant
to impact penetration, which could be encountered when back filling, or when
subsequent soil movements occur than the softer types of protective pipe
coatings. Further, as cement cures under water it may be used in places where
pipes are partly submerged in water without the need for draining of trenches
before application of the tape. This is an advantage over some conventional
protective coverings which require a dry surface for application and bonding.

(2) Insulated Pipes

With the great emphasis now being placed on the efficient lagging of pipes,
a number of insulating materials are being more widely used. Typical of these
materials are calcium silicate, mineral wool, polyurethane foam and expanded
polystyrene, which are fragile and are easily damaged through impact. This can
lead to the loss of insulation value and in the case of the asbestos used in
older installations, a serious health hazard can result. Furthermore some of
the insulants may be flammable and so will require an additional fire protective
over-layer. In the case of non-flammable materials any protective coverings
must not detract from the overall fire resistance. Cement tape product has been
subjected to fire tests to BS 476 and the interpretation of the results shows
that the product can be applied to flammable insulants to provide an upgrading
in their fire resistance while when used in conjunction with a non-flammable
insulant it will not detract from that system's fire rating. Insulants overlaid
with cement tape can be given improved impact resistance due to the hard, resilient casing. We have mentioned earlier that impact damage to asbestos insulation renders the surface friable with a consequent health hazard.

Fortunately asbestos insulation materials are no longer in large usage due to
the production of more efficient forms of insulation. On older pipe installations, however, it is quite common to find that asbestos has been used. Due
to factors such as age, mechanical damage and subsequent pipe repairs, the

surface of the asbestos becomes friable and not only presents a health hazard
in its own right but removal is also hazardous. This process requires the use
of specialist equipment and expertise in handling and it may result in areas
having to be isolated while the asbestos is removed. Special provisions are
also necessary for disposal of old asbestos insulation. In certain areas the
use of cement tape can provide a simple solution to the problem. Its effect
is two-fold in that during application over the asbestos insulation a fine paste
of cement penetrates the surface and binds the loose fibres of asbestos together
whilst, once cured, it also provides a hard impermiable casing to the insulation.

Earlier in this paper it was shown that the cement-tape product is not a vapour
barrier, i.e. it allows water vapour to pass through it. This can be of great
advantage when used over exposed, porous insulating materials. Porous insulants
if left exposed to the weather will absorb water and if vapour and waterproof
membranes are applied whilst water is present, the subsequent development of
vapour pressure between the insulant and the vapour-proof barrier may cause
rupture in the protective layer. The tape product, on the other hand, can be
applied directly to wet substrates to provide them with an almost immediate
weather-proof covering, but which will allow the moisture to evaporate through
it. No vapour will thus build up within the insulant to cause delamination and
blistering of the protective layer. The system is thus "self-venting" and after
sufficient time has elapsed to allow the insulation to dry out, a bituminous or
alkyd paint may be applied to render the surface more waterproof or to serve as
a decorative finish to the surface.

The cement tape product may be used in conjunction with many conventional insulation systems, and in comparison with normal finishing schemes can show
significant advantage.

Insulation	Normal finishing scheme	Cement tape finishing scheme
12.5 mm cellular polystyrene	Hard setting cement + canvas + paint finish	1 layer tape + paint finish
50 mm cork	12.5 mm wire netting, 12.5 mm hardsetting cement + canvas + paint finish	1 layer tape + paint finish
Preformed P.U. cellular polyurethane	12.5 mm wire netting, 12.5 mm Keene's cement + paint finish	1 layer tape + paint finish
Silicate sections	Proprietary hard setting cement	1 layer tape + paint finish
Cork slab	Fixing bands for slab + 2 layers felt + wire netting + bituminous paint	1 layer tape + paint finish

As can be seen from the above example specifications, at
least one and in some cases two, separate stages of a
typical system can be removed from the total materials and
man hour costs.

Summary

A cement/cotton tape has been developed for use in the pipe industry to
provide a corrosion protection to steel pipes or as a protective layer to
pipe insulation materials. When cured it provides an impact-and weather-
resistant coating to the pipework. It is a material which can be applied

without the need for skilled labour or sophisticated application equipment and
which can provide a cost effective system in certain types of use.   Its
introduction to the pipe industry is relatively recent and certain areas in
its application are presently being investigated by ourselves and a number of
companies established within the pipe industry and allied organisations.
The product is marketed by Evode Limited under the trade-name 'ROK-RAP'.

ACKNOWLEDGEMENTS

The authors acknowledge the work of the research and development group,
Drexel University, Philadelphia, Pennsylvania, USA, in providing test data
on ROK-RAP.   They also wish to thank the Directors of Evode Limited for
permission to publish this paper.

REQUIREMENTS OF ANTICORROSIVE COATING

1)   GOOD ADHESION TO PIPE SURFACE

2)   MINIMUM SURFACE PREPARATION

3)   COATING IN EASILY HANDLABLE FORM

4)   MINIMUM APPLICATION EQUIPMENT

5)   VERSATILITY TO COVER PIPE SECTIONS, JOINTS, REPAIR

6)   GOOD CHEMICAL, MECHANICAL AND THERMAL RESISTANCE

7)   LOW COST

8)   RELIABLE

REQUIREMENTS OF INSULANT PROTECTIVE COATING

1)   GOOD ADHESION TO VARIETY OF INSULATION MATERIALS

2)   EASE OF APPLICATION

3)   GOOD MECHANICAL RESISTANCE PARTICULARLY TO IMPACT

4)   GOOD THERMAL RESISTANCE

5)   FIRE RESISTANCE

6)   GOOD AESTHETIC APPEARANCE

7)   CAN BE COLOUR CODED E.G. PAINTED

8)   VAPOUR BARRIER FOR SUBAMBIENT TEMPERATURES

<u>PROPERTIES OF ROK-RAP</u>

SURFACE PREPARATION	FREE FROM CONTAMINANTS
	CLEAN
	NEED NOT BE DRY, WET AN ADVANTAGE
PRIMER/ADHESION PROMOTER	INCORPORATED IN TAPE
PHYSICAL FORM	SINGLE COMPONENT TAPE
APPLICATION	ACTIVATION OF TAPE WITH AQUEOUS MEDIUM
	SINGLE/MULTIPLE LAYERS
	SPIRAL/CIGARETTE WRAP
	FACTORY/SITE/REMEDIAL REPAIR
APPEARANCE	GREY COATING
IMPACT RESISTANCE	VERY GOOD
THERMAL RESISTANCE	$- 160^{\circ}C \longrightarrow + 150^{\circ}C$
CHEMICAL RESISTANCE	AFFECTED BY STRONG ACIDS
	GOOD SULPHATE RESISTANCE
	GOOD RESISTANCE TO ORGANIC SOLVENTS

<u>CONVENTIONAL PROTECTIVE SYSTEMS</u>

	MOLTEN SYSTEM	TAPE	PLASTIC FILM	POLYMERIC POWDER
EXAMPLE	COAL TAR/PITCH	PRESSURE SENSITIVE	POLYETHYLENE	EPOXIDE
APPLICATION	MOLTEN WITH MEMBRANE	SPIRAL/ CIGARETTE WRAP	ADHESIVE WRAPPED	POWDER COATING WITH FUSION
PRIMER	YES	YES	NO	NO
SURFACE PREPARATION	CLEAN, DRY PREHEAT PIPE	CLEAN, DRY	SECONDARY COATING	CLEAN, DRY
PRODUCTION	YES	YES	YES	YES
SITE	YES	YES	DIFFICULT	DIFFICULT
REMEDIAL/ REPAIR	YES	YES		
ADVANTAGES	LOW COST	EASY APPLICATION	PROTECTIVE COATING	BETTER PHYSICAL PROPS
PROBLEMS	PIN HOLING BRITTLENESS THERMAL LIMITS	INTERFACIAL ADHESIVE THERMAL LIMITS	APPLICATION CONDITIONS	UNIFORMITY OF FILM THICKNESS/ COMPOSITION PIN HOLING ADHESION BRITTLENESS

fig 1.1
ROK RAP CURING TESTS
COMPRESSION
TENSION
STRENGTH — MNm⁻²
DAYS CURING
fig 1.2
CURING TESTS - ULTIMATE STRENGTH
ROK RAP
MORTAR
STRENGTH — MNm⁻²
COMPRESSION
TENSION
DAYS CURING

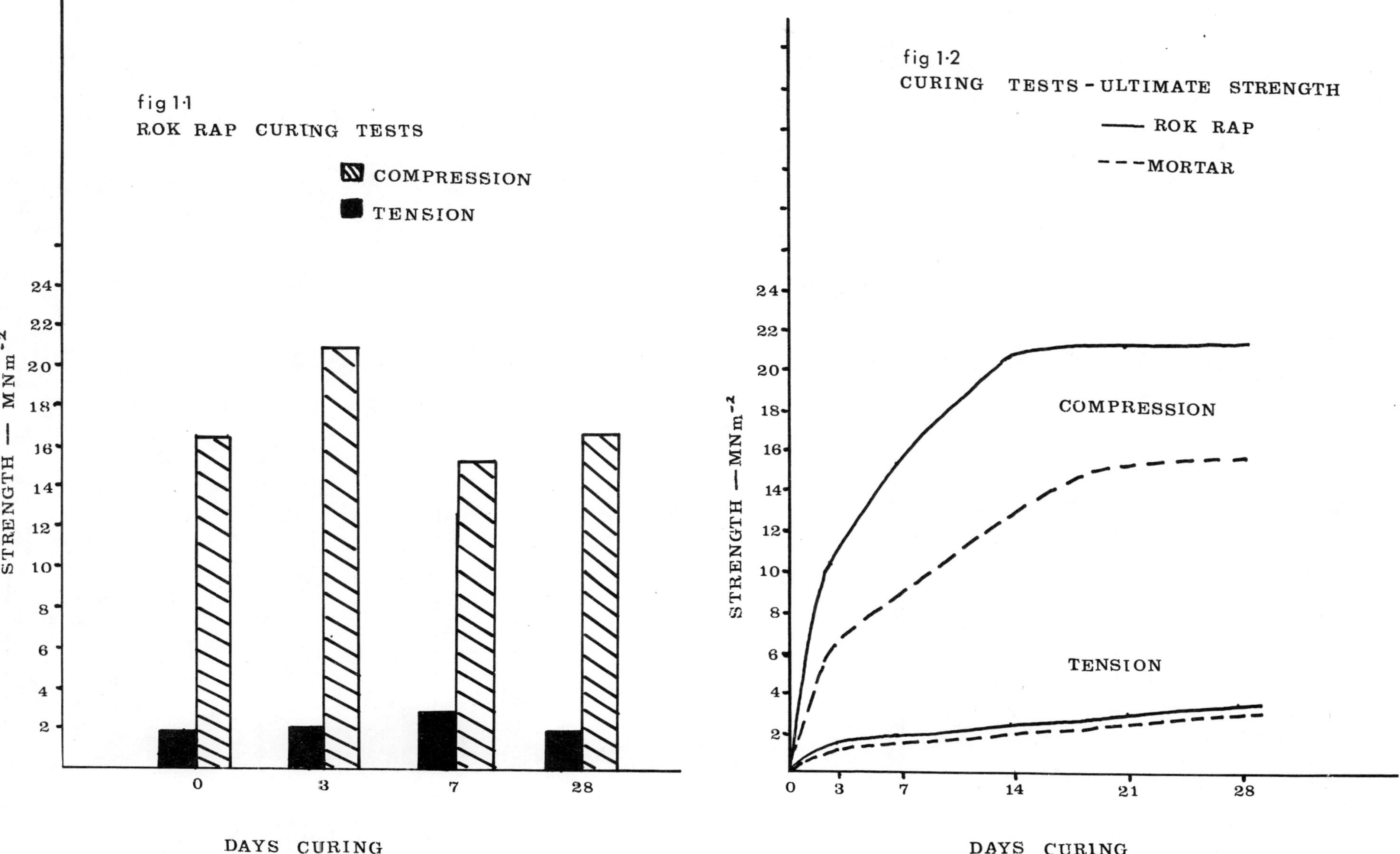

105

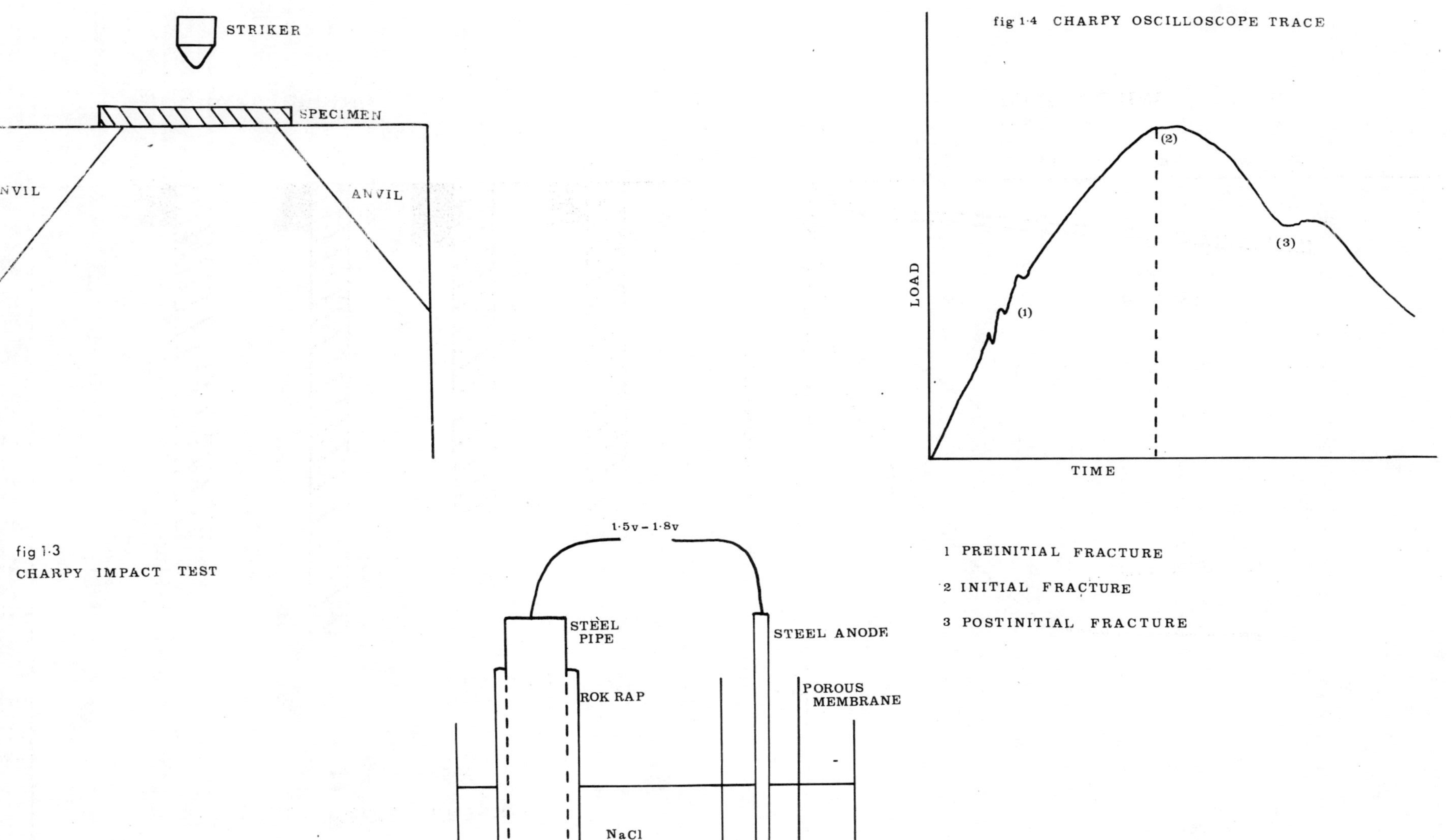

STRIKER
SPECIMEN
ANVIL
ANVIL
fig 1·3
CHARPY IMPACT TEST
fig 1·4 CHARPY OSCILLOSCOPE TRACE
LOAD
TIME
(1)
(2)
(3)
1 PREINITIAL FRACTURE
2 INITIAL FRACTURE
3 POSTINITIAL FRACTURE
1·5v - 1·8v
STEEL PIPE
ROK RAP
STEEL ANODE
POROUS MEMBRANE
NaCl SOLUTION
fig 1·5   CELL DIAGRAM

POLYETHYLENE COATED PIPES FOR SUBMARINE PIPELINE

M. Tanaka, F. Otsuki, F. Hirano and T. Sato

Nippon Steel Corporation, Japan

Summary

Polyethylene coated pipes, concrete jacket pipes and non-concrete jacket pipes were developed and their properties were individually investigated. Firstly voids formation between pipe surface and polyethylene and the surface structure of coated polyethylene concerned with the coating process is discussed. Secondly, the mechanism of grip strength of concrete jacket to polyethylene coated pipe is elucidated with polyethylene coated pipes processed by an annular die and a flat die. Thirdly the importance of adhesive between polyethylene and pipe is made clear in a wide temperature range. Finally effects of surface structure, damages and degradation to environmental stress cracking resistance were investigated in detail with various kinds of polyethylenes. Summarizing all these results, important points and properties required for polyethylene coating used for deep sea pipeline are elucidated.

Held at Imperial College, London, England.
Organised and Sponsored by BHRA Fluid Engineering

1. INTRODUCTION

Recently the construction of offshore pipeline systems, such as the trans-Mediterranean gas pipeline or the North Sea crude oil pipeline, has been gradually on the increase. Polyethylene coated steel pipes which were developed and commercially manufactured by us were first laid in 1976 for the offshore pipeline system in the bay of Tokyo (Ref. 1). After that, the new type polyethylene coated pipes which maintain high and stable performance over a long term in a wide temperature range of from $-60^{o}C$ to $+80^{o}C$ were developed (Ref. 2). This time, polyethylene coated pipes for the deep sea (over 300 meters) pipeline system, some part of which is covered by concrete jacket to stabilize the pipeline on the sea bottom against the buoyancy of sea water have been developed, making use of our various experiences. The concrete jacket is not only useful for the weight, but is also useful to protect the polyethylene coating from damage from external forces.

The following two characteristics of polyethylene coated pipes with concrete jacket for the deep sea pipeline are requested in comparison with pipes for the shallow sea pipeline. Firstly pipes, particularly the concrete jacket and the polyethylene **coating**, must resist the tremendous compressive force to prevent pipe sagbend by the tensioning machine during construction. That is, it is necessary that (1) the concrete jacket doesn't slip off the polyethylene coated steel pipe in the limit of concrete jacket destruction (2) the polyethylene coat is not destroyed or deformed and (3) the polethylene coat doesn't slip off the steel pipe caused by the tremendous compressive force from the tension gripper and the weight of the constructing pipeline in the sea.

Secondly the coating polyethylene must possess high resistibility to environmental stress cracking (ESCR). Deepsea pipelines involve thicker pipe walls to withstand bending or buckling collapse failures. In the end, it should be more economical to lay pipe without weight coating relying on the polyethylene coated pipe alone to provide sufficient submerged weight for on-bottom stability.

When the polyethylene coated pipe without concrete jacket is compressed with tremendous force and laid down in the sea from the stinger, it cannot be helped that to some extent the surface of the polyethylene layer is damaged. These damages, scratches or dents, make the ESCR of the polyethylene material abruptly worse. The excellent chemical and physical properties of polyethylene ensured the rapid progress of the quality level of the anticorrosive to the pipeline. Polyethylene coating systems for pipeline have the advantage of using the convenient impaction method for concrete placing using the resilient property of thicker polyethylene coating layer and the advantage of cathodic protection very effectively using the superior insulation property and the pinhole-free layer. The environmental stress cracking resistance (ESCR) is, however, a unique weak point; this phenomenon is well known on the polyethylene sheath for submarine communication cable (Ref. 3). Though the ESCR of polyethylene depends on the level of the inner stress, the presence of scratches or dents on the surface of the material accelerates drastically the lowering of ESCR under an inner stress. As the dependency of the acceleration to lower the ESCR differs with different types of polyethylene; the polyethylene which has the superior ESCR must be selected as the anticorrosive material used for submarine pipeline.

In this paper the qualities demanded for the deep sea pipeline system and the actual qualities of polyethylene coating pipes developed by us are to be discussed.

2. POLYETHYLENE COATING

2.1. Polyethylene coating process

Pipe coating systems with polyethylene can be classified into a powder melt coating system and an extrusion coating system, then the latter can be classified into an annular die extrusion coating and a flat die extrusion coating according to their extrusion die shape (Ref. 4). The flat die extrusion coating system is commonly used for large diameter pipes. The manufacturing flow sheet and the process of the polyethylene coating system are illustrated in Figures 1 and 2 respectively. A preheated steel pipe (about 200°C) cleaned by blast before hand is coated by adhesive and polyethylene. The molten polyethylene sheet extruded horizontally from a flat die is helically wrapped around the rolling pipe with advancing by skew rollers. The degree of the overlapping part of the polyethylene sheet on the pipe depends on the ratio of the width of polyethylene sheet and the advanced distance of the pipe during one revolution. The coating thickness can be controlled by the number of overlapping layers and the thickness of the sheet decided by the ratio of extrusion rate and the revolution rate of the pipe (draw down ratio). It is a pecular merit of this coating system that various size pipes can be coated with the desirable thickness layer. On the interface of polyethylene sheets of the wrapping part, two polyethylene sheets are perfectly unified by strong contact at a high temperature molten state. On the coating surface a seam line corresponding to the edge of the polyethylene sheet can be spirally observed.

2.2. Correlation of the cross-sectional shape of polyethylene sheet and the appearance of the coated pipe.

When coating by the powder melt coating system or the annular die extrusion coating system mentioned above the coating surface is essentially flat and uniform. On the other hand in coating by the flat die extrusion system, the helical seam line and slight undulations remain on the coating surface. The shape of the seam line and the undulation is very important for the non-conrete jacket pipe from the view point of appearance and mechanical properties of the polyethylene coat.

Various cross-sectional shapes of wrapping polyethylene sheets (A in Figures) and the way of overlapping on the pipe (B in Figures) are illustrated in Figures 3 and 4.

When the uniform thickness sheet is helically wrapped on a pipe with a helical pitch of half width of the sheet, voids formation at the interface of polyethylene and the pipe and linear sharp protrusions along the seam line are problems. For solutions of these problems and to save polyethylene, the special cross-sectional shaped polyethylene sheets of which both ends are gradually thinner as shown A in Fig. 3 are used. The special extrusion die and polyethylene having proper rheological properties in molten state, melt extensibility and melt extension, are requested in order to obtain steadily those shaped polyethylene sheets (Ref. 5). Smoother coating surface even at seam line can be formed and voids do not generate interface of polyethylene and pipe using these shaped polyethylene sheet. As the turning rate or advancing rate of the pipe is not constant caused by the slight oblates of the circle shape and the straightening of the pipe or caused by skidding of skew rollers, the spiral pitch of the over-lapping polyethylene sheet fluctuates a little and bigger protrusions (b in Figure 3) than normal size protrusions (a, c, e in Figure 3) or dints (d in Figure 3) are generated. These fluctuations are not desirable from the appearance point of view and from the possibility of falling short of the lowest required thickness at dints. Still more the fluctuation of coated polyethylene thickness using these shaped sheets caused by the shift of the overlapping position is smaller than that using uniform thickness sheet. Usually polyethylene sheet is overlapped by three layers near the seam line, therefore thinner seam lines like d in Figure 3-B are scarcely formed.

On the contrary, some degree of polyethylene surface undulation is necessary for the concrete jacket pipe for the increase of frictional resistance between polyethylene and concrete. As polyethylene is not chemically active to concrete material on adhesive, several points must be considered to increase the frictional resistance between polyethylene and concrete, for example use of a rubber sheet on the polyethylene coat or to put ruggedness on the surface of polyethylene coat (Ref. 6). When the flat die extrusion system is adopted, the surface undulation can be produced at one's will using specially designed cross-sectional shaped sheets, as shown in Figure 4.

3. FRICTIONAL RESISTANCE BETWEEN POLYETHYLENE COATING AND CONCRETE JACKET

3.1. Preliminary test

Frictional resistances between concrete jacket and polyethylene coated pipes of 203.2 mm diameter which has a smooth surface coated by an annular die and having a spiral protrusion seam on the surface coated by a flat die were measured. The height of protrusion changed from 0.25 mm to 1.25 mm with 0.5 mm interval. The concrete jacket was placed by casting.

The thickness of the concrete jacket of a test specimen which has just one circle of spiral seam is 40 mm and its width is 100 mm. The pipe was vertically set and supported to the concrete jacket by a backing plate; the steel pipe was then pushed down by the test machine. The applied force to push down the pipe at the beginning of pipe movement was divided by the total interface area between polyethylene and concrete; then the value was defined to shear adhesion strength. As the cross-sectional shape of protrusion at the seam is not symmetrical with respect to the pipe direction, pipes were pushed down from two opposite directions, A and B as indicated in Figure 5.

Shear adhesion strength depends on the height of protrusion in both cases, A and B. The shear adhesion strength of a test pipe with a 0.8 mm high protrusion on the polyethylene coat is double that without a protrusion. This result means that protrusions of polyethylene act as anchors to resist the slip between polyethylene and concrete.

3.2. Tests by practical use pipe

The concrete placing method in which concrete with high speed is continuously sprayed to the pipe surface, the so called impingement method, is one of the best methods to jacket concrete on the pipe. Frictional resistance between polyethylene and concrete or various pipes of 609.6 mm diameter coated by a flat die and an annular die and concrete and were placed by the impingement method were measured. On polyethylene surfaces of pipe coated by an annular die, no distortion could be observed and on that coated by a flat die, spiral protrusion could be observed. The height of the protrusion fluctuated between 0.65 mm and 1.05 mm as shown in Figure 6 and its average height was 0.85 mm. The 80 mm thick concrete jacket includes two coaxial circular wire meshes and physical parameters of the concrete are shown in Figure 7. Test specimens were cut as shown in Figure 8, where just one circle of spiral protrusion remains on the polyethylene surface. The test method is shown in Figure 8, where the stress detector was set on the backing plate and the strain sensor was attached on the pipe surface. One typical test result is shown in Figure 9. The shear adhesion strength (Sa) is defined by the stress at the commencement of slip moving of the pipe (Pa) divided by the interface area (S). The maximum friction strength (Sf) is defined as the maximum stress (Pf max) divided by S.

Behaviour of frictional resistances of two pipes coated by an annular die and a flat die are very different from each other. In the case of a pipe coated by a flat die, stress reaches a maximum at the beginning of slip and gradually reduces. On the other hand, in the case of pipe coated by an annular die, the stress is very small at the beginning of slip and increases gradually. The maximum stress displacement (r) is defined by the ratio of the amount of slip (r_p) at the maximum stress to the total length of the concrete jacket (L). The shear adhesion strength (Sa) is linearly related to the maximum stress displacement (r_p) on a log scale in both cases coated by two different types of die as shown in Figure 10. In practical use the pipe which has a larger shear adhesion strength and smaller maximum stress displacement is desirable. The shear adhesion strength (Sa) is also related to the maximum friction strength (Sf) for both cases of different coating systems as shown in Figure 11.

It is concluded from these experimental results that the pipe coated polyethylene done by the flat die is better for the concrete jacket pipe than that done by the annular die. As the concrete jacket pipe coated polyethylene by the flat die of which the shear adhesion strength can reach $2kg/cm^2$, a pipe with a diameter of 610 mm endures 38 ton force applied from outside per one metre length.

4. RESISTANCE OF POLYETHYLENE TO COMPRESSION

In the case of concrete jacket pipes the compression stress to polyethylene coating is comparatively small owing to the concrete jacket. On the other hand, in the case of the non-concrete jacket pipe the compression force from the tension gripper acts directly on the polyethylene coat and it is requested that the polyethylene coat is never destroyed or deformed due to the compression force.

Yield stresses and elastic moduli on compressive deformation of three kinds of carbon blended (2.5 wt.%) polyethylenes, high pressure process polyethylene (density; 0.935), two middle or low pressure process polyethylenes (density 0.943 and 0.965) in the temperature range of $-60^{\circ}C$ to $30^{\circ}C$ were measured as shown in Figures 12 and 13. Even on the high pressure process polyethylene at $30^{\circ}C$ the yield stress is larger than 100 kg/cm^2 and the elastic modulus is larger than 1,500 kg/cm^2; these values are enough for practical use. When the concrete jacket pipe was squeezed flat and the concrete jacket was broken, a kind of damage could be observed on the polyethylene coating of the deformed pipe.

5. SLIP BETWEEN POLYETHYLENE COATING AND STEEL PIPE

The slip between polyethylene coating and the steel pipe must not occur by the applied tension during laying. The adhesive between polyethylene and pipe surface depends on the pipe surface condition and adhesion material. Adesive materials used in practise are classified into two categories, hard adhesive and soft adhesive.

Though copolymers of ethylene and a second component possessing an adhesive property, vinyl acetate, acryl acid or metaacryl acid, belonging to the hard type of adhesive must be treated under severe conditions in the coating process, the change of adhesion strength in the wide temperature range is comparatively small.
On the other hand adhesion strength of adhesive blended with bitumen, synthetic rubber and wax belonging to the soft type of adhesive depends sensitively on temperature. Temperature dependencies of adhesion strength of these two types of adhesive are shown in Figure 14. The adhesion strength of the hard adhesive comes mainly from chemical bonds between steel and adhesive material and that of the soft adhesive comes mainly from their glutionous properties which relate deeply to the viscosity. At a high temperature, over $0^{\circ}C$, the adhesion strength of the soft adhesive depends on its viscosity which decreases drastically with temperature increase, and at low temperatures below $0^{\circ}C$, the adhesive becomes brittle because of a lower temperature than its glass transition temperature, which reduces the impact resistance of the whole polyethylene coating layer (Ref. 2). The adhesion strength of the hard adhesive at $60^{\circ}C$ is, however, still higher than 40 kg/cm^2, which is ten times stronger than the frictional strength between concrete jacket and polyethylene coating surface. The adhesion strength of the soft adhesive is 50 kg/cm^2 at $0^{\circ}C$ and 1 kg/cm^2 at $60^{\circ}C$. Considering the situation that the temperature of a pipe without a concrete jacket reaches $70^{\circ}C$ by sunlight in summer, the adhesion strength at $60^{\circ}C$ is too low for polyethylene coating to be prevented from slipping on the pipe surface.
The adhesive consisting of polymer material is viscoelastic and it is expected that the adhesion strength will change with the shear rate according to the theory. The shear rate dependencies of adhesion strength were measured as shown in Fig. 15. The dependency of the soft adhesive is larger than that of the hard adhesive.
The smaller dependency is desirable in practical use.

6. ENVIRONMENTAL STRESS CRACKING RESISTANCE (ESCR) OF POLYETHYLENE

6.1. Effect of protrusion on the polyethylene surface ESCR

The resistance of polyethylene to cracking is made catastrophically worse by the existence of defects, notches or scratches. As mentioned above it is endeavoured to make the polyethylene coating surface flat for the non-concrete jacket pipe. However an uneven surface is requested for the concrete jacket pipe to enable frictional resistance between polyethylene and concrete. The effect of steps on the polyethylene sheet surface to ESCR was tested with five different types of polyethylene as shown in Table 1, where test conditions and descriptions are mentioned. Along the centre line of the test specimen three different steps, 1.0, 0.5, 0 mm high were made as indicated in Table 1.

All kinds of polyethylene samples without steps (step height is 0 mm) except for one middle or low pressure process polyethylene did not fail until after 300 hrs. immersion in Igepal solution. It is clear that the existance of a 0.5 mm high step on the surface makes ESCR worse in middle or low pressure process polyethylenes and the existance of 1.0 mm high step makes ESCR worse in all kinds of polyethylene except copolymer of high pressure process polyethylene. In conclusion, uneven surfaces in polyethylene coating layer is required on the concrete jacket pipe for frictional resistance between polyethylene and concrete; the uneven surface makes ESCR of polyethylene worse except copolymers of high pressure process polyethylene.

6.2. Effect of mechanical mixture of molten polyethylene to ESCR

As described above polyethylene surfaces even-coated by a flat die for non-concrete pipes is made flat using a singular sectional shaped polyethylene sheet in consideration of ESCR. However it cannot be avoided that some damage such as bruises, scratches or abrasions is subjected to the polyethylene coating surface during transportation or during laying. These damages make ESCR of polyethylene drastically worse. In addition to that the polyethylene coat of non-concrete pipes is directly exposed to sunlight, heat or other influences of the environment and then the process of degradation, is accelerated, which makes ESCR of polyethylene catastrophically worse with synergetic effect of damages.

Degradation of polyethylenes were brought by mechanical mixture in molten state at around $200^{\circ}C$ to test the effect of degradation to ESCR. According to the description of ASTM D1693, ESCR of 0.5 mm depth notched polyethylene specimens, were measured. Three kinds of polyethylene with various mixture times were measured as shown in Figure 16. Times of 50% failed original polyethylenes, copolymer of middle or low pressure process, homopolymer and copolymer of high pressure process, are 7 hrs, 80 hrs and over 300 hrs in turn. The times of the initial two polyethylenes decrease radically with mixture time, but the time of copolymer of high pressure process polyethylene does not decrease with mixture time. These results mean that a copolymer of high pressure process polyethylene possesses excellent properties of ESCR against surface damages or degradation.

7. CONCLUSIONS

Steel pipe coated by polyethylene which retains superior chemical and physical properties is extremely valuable for protection from external corrosion; this being the most important problem for the safety of pipeline and they are widely used throughout the world. The following two properties are taken seriously from the view of deep sea pipeline. Using the polyethylene coated pipes for deep sea pipeline, the resistibility to the enormous compression by tensioner and environmental stress cracking resistibility (ESCR) are of added importance. For the production of polyethylene coated pipes to satisfy the required condition, the polyethylene coating process by flat die was improved and a new type of polyethylene was developed. For the frictional resistance between polyethylene and concrete, proper sizes of spiral protrusion line were made on the polyethylene coating surface. For the adhesion strength between polyethylene and the pipe surface, the hard adhesive was used. Finally for ESCR and degradation, the copolymer of high pressure process polyethylene was selected.

REFERENCES

1) Michishita, T., Tsuchida, I., Kawakami, M., and Sudo, T., "Laying of a submarine pipeline with plastic coating under Tokyo Bay". 2nd International Conference on the Internal and External Protection of Pipes. Paper C2. Orgaised by BHRA Fluid Engineering. Canterbury England (Sept., 7th - 9th, 1977).

2) Tanaka, M., Otsuki, F., and Sugimura, S., "The performance of polyethylene coated pipe for arctic service". Corrosion '79 The National Association of Corrosion Engineers Annual Conferences. Paper 260. Atlanta, U.S.A., (March 12th - 16th, 1979).

3) DeCoste, J.B., Malm, F.S., and Wallder, V.T., "Cracking of stressed polyethylene".
Ind., and Engineering Chem., *43*, 1 pp117 - 121 (Jan., 1951).

4) Hauk, V., Meyer, W., and Schwenk, W., "Required coating thickness for external
protection of buried line pipe and polyethylene". 1st International Conference on the Internal
and External Protection of Pipes. Paper B2. Organised by BHRA Fluid Engineering.
England (Sept., 9th - 11th, 1975).

5) Tanaka, M., Otsuki, F., Kinugasa, J., Nagasawa, T., and Tsurutani, I., "Rheological
properties of polyethylene in the pipe coating process". Society of Plastics Engineers,
37th Annual Technical Conference. Paper 128 - 29. New Orleans, U.S.A. (May 7th - 10th,
1979).

6) Ayusawa, S., Tanaka, M., Yoshida, H., Otsuki, F., "Polyethylene coated pipe with rugged
surface to improvement the friction resistance between the concrete-jacket and the
polyethylene coating". Japan patent publication No. 54-3699 (Feb., 26th 1979) (in Japanese).

Polyethylene _density Melt Index_	High Pressure Process PE		Middle or Low Pressure Process PE		
	Homo-polymer	Co-polymer	Co-polymer	Co-polymer	Homo-polymer
Step height, h (mm)	0.931	0.935	0.943	0.953	0.965
	0.12	0.13	0.17	0.38	0.50
0	NF	NF	NF	NF	4.1
0.5	NF	NF	99.0	35.0	3.4
1.0	24.0	NF	41.0	34.0	2.0

Annotation : 1) Density
 Including carbon black 2.5%
2) Test piece

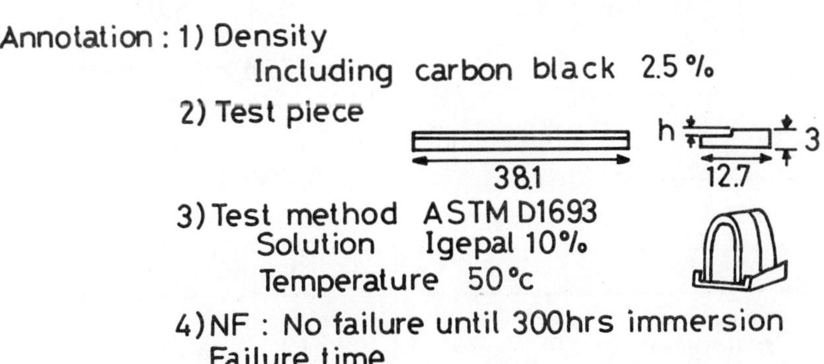

3) Test method ASTM D1693
 Solution Igepal 10%
 Temperature 50°c
4) NF : No failure until 300hrs immersion
 Failure time
 Time till 50% of test pieces failed

Table 1. Environmental stress cracking resistance of various
polyethylenes by samples with protrusion on the surface.

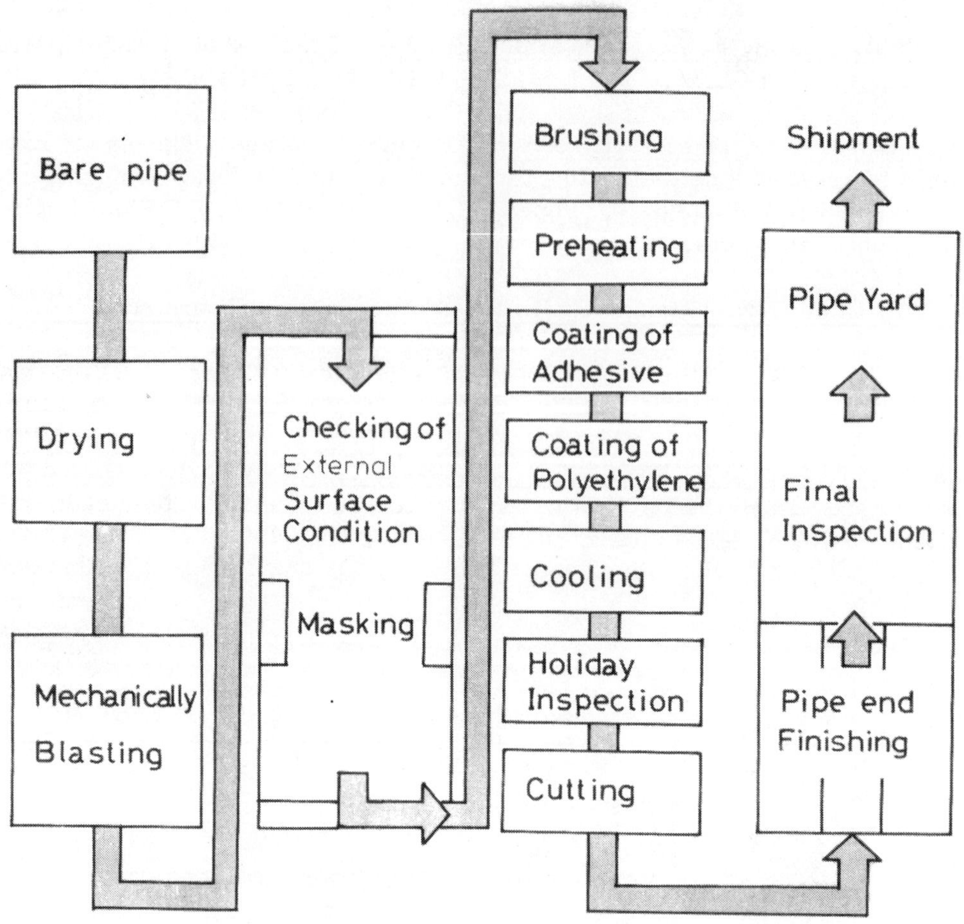

Fig. 1. Manufacturing flow sheet for polyethylene coated large diameter steel pipes.

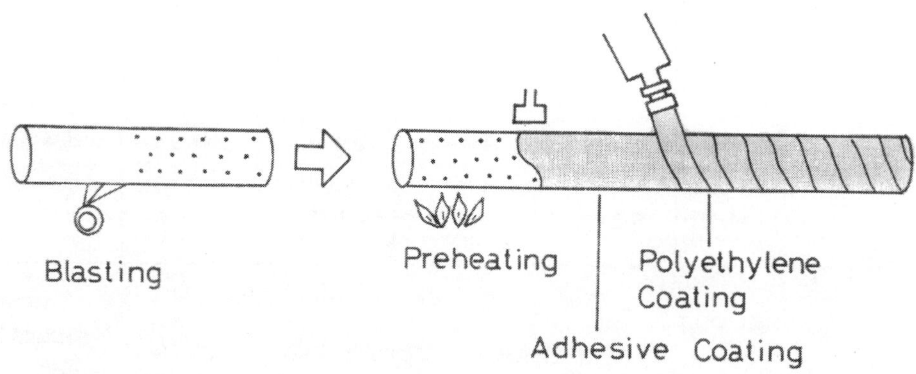

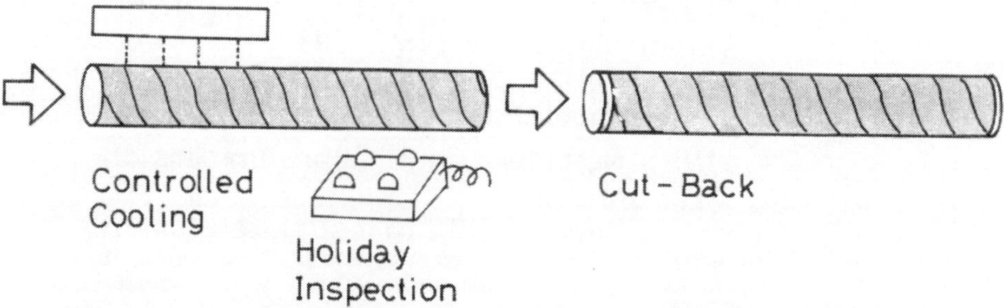

Fig. 2. Polyethylene coating process for large diameter steel pipes by the flat-die extrusion system.

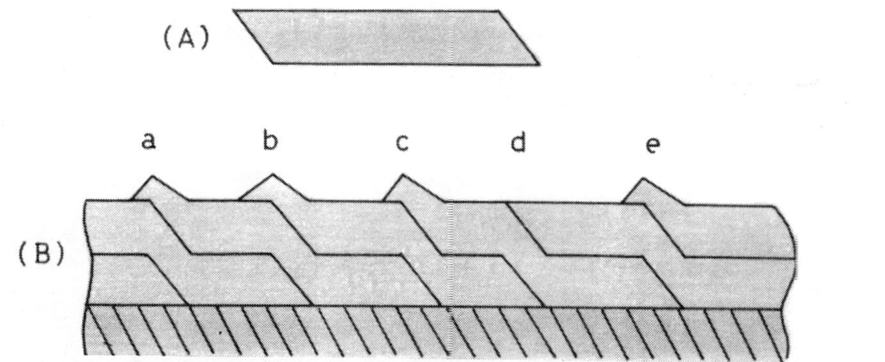

Fig. 3 Schematical cross-section view (A) of both ends of thinner polyethylene sheet (B) wrapped helically on a steel pipe.
a, c, e; normal seams   b, d; abnormal seams

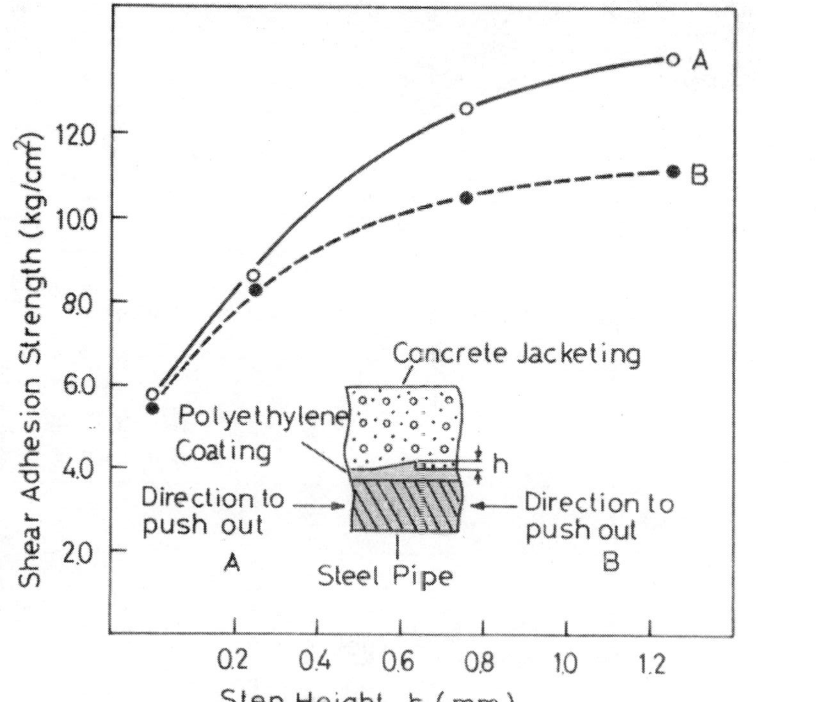

Fig. 5 Grip strength (shear adhesion strength) of the concrete jacket to polyethylene coated steel pipe with various step heights (h) on the surface of the coating.

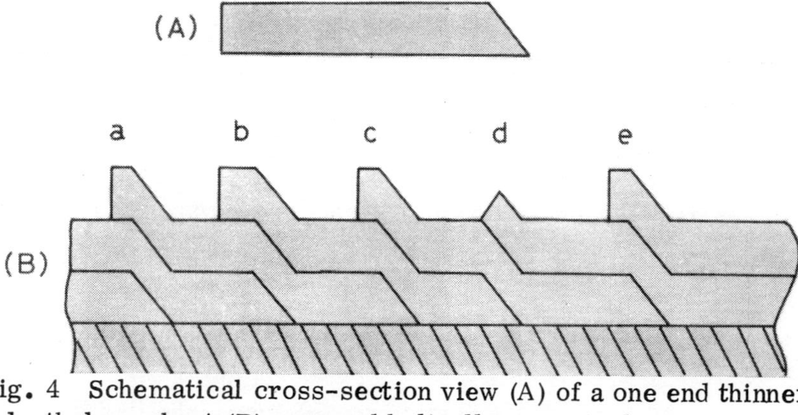

Fig. 4 Schematical cross-section view (A) of a one end thinner polyethylene sheet (B) wrapped helically on a steel pipe.
a, c, e; normal seams   b, d; abnormal seams

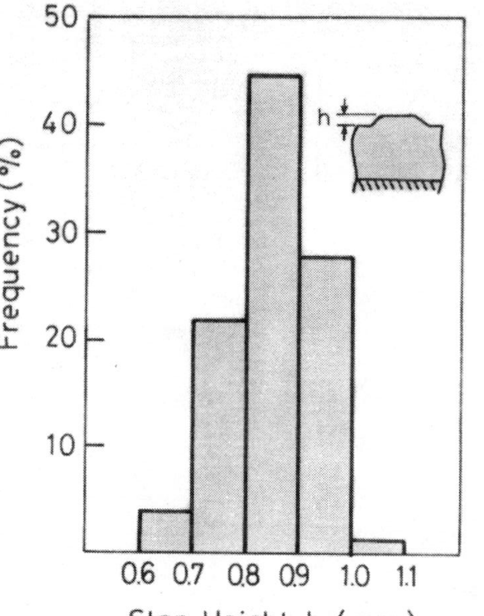

Fig. 6 Distribution of step heights on the polyethylene coating surface.

Polyethylene Coating	Concrete Jacketing		Wire Mesh
	Density ※	Compressive Strength ※	
Flat–die Extrusion Coating	2.22g/cm^3	341 kg/cm^2	Galvanized Steel Wire Mesh Diameter 1.4mm Hexagonal Pattern 1.5x1.5 in
Annular–die Extrusion Coating	2.31	330	

※ Data at 4weeks aging after concrete placing

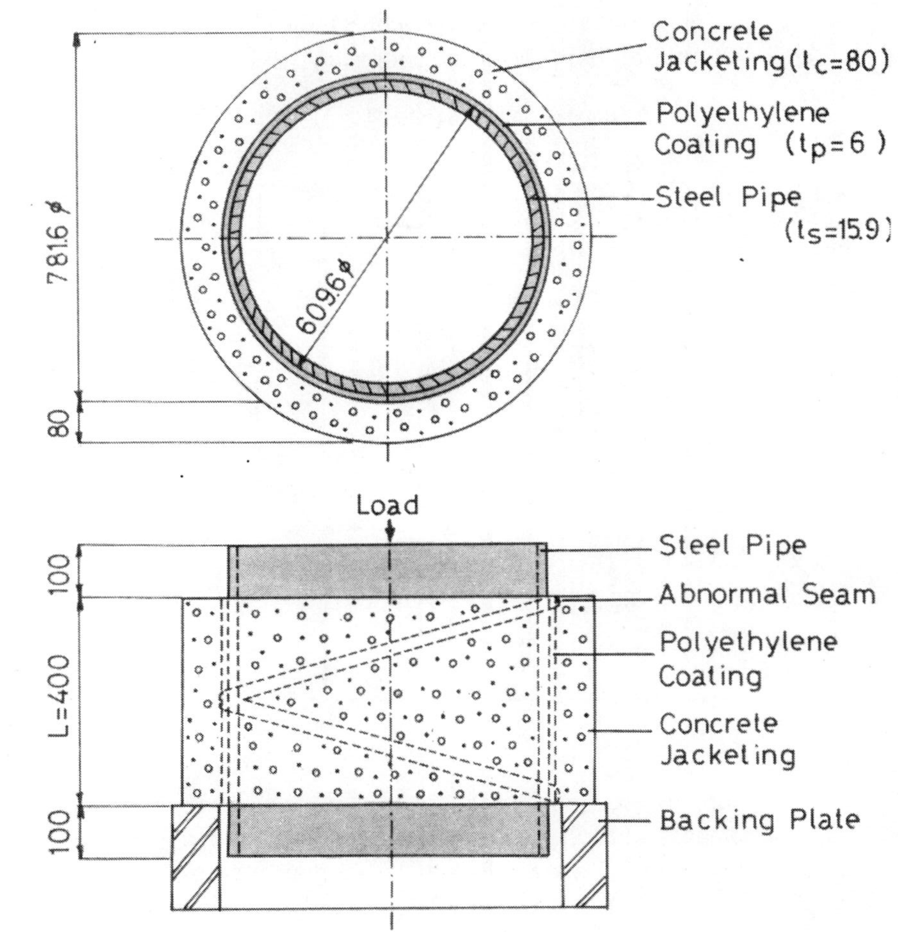

Fig. 7. Constitution of a concrete jacketing pipe.

Fig. 8. Test specimen to measure the shear adhesion strength

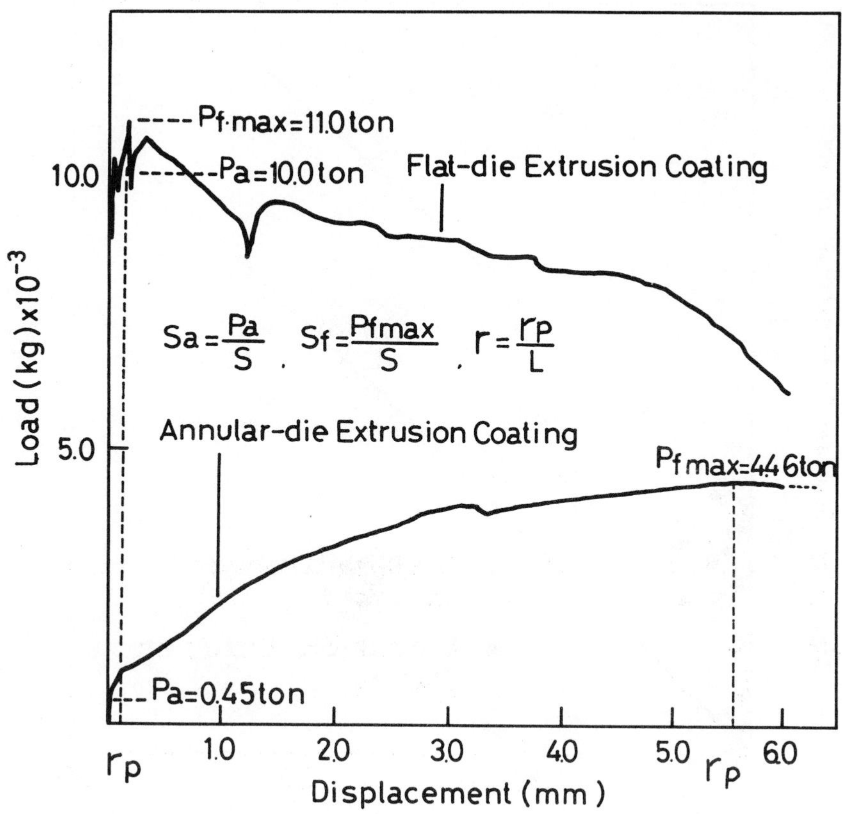

Fig. 9. Schematical diagram of the vertical push out test

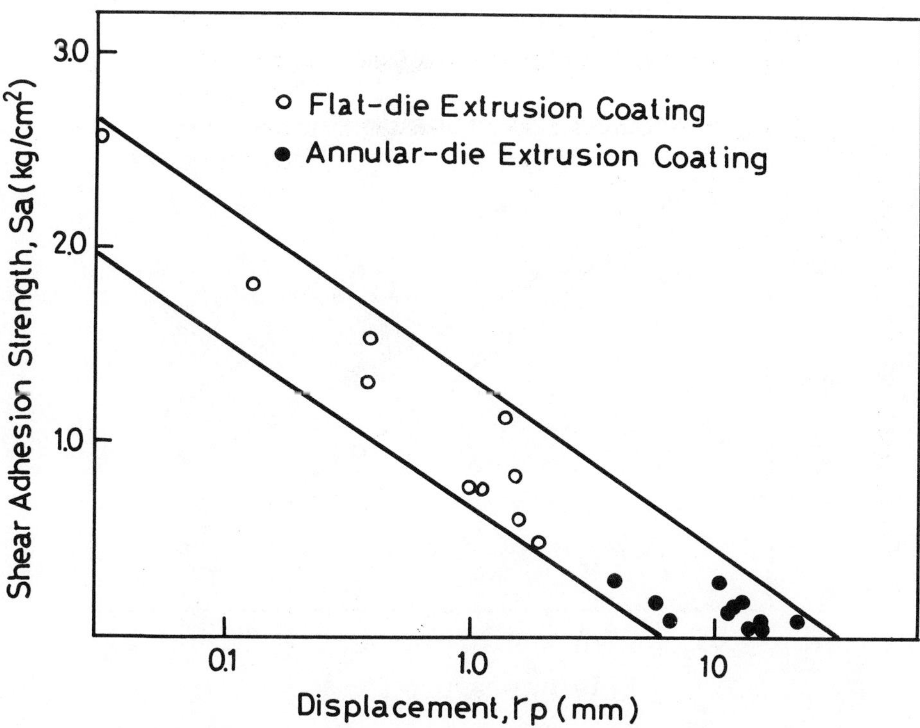

Fig. 10. Variations of push out forces with displacements of pipes to concrete jacketings.

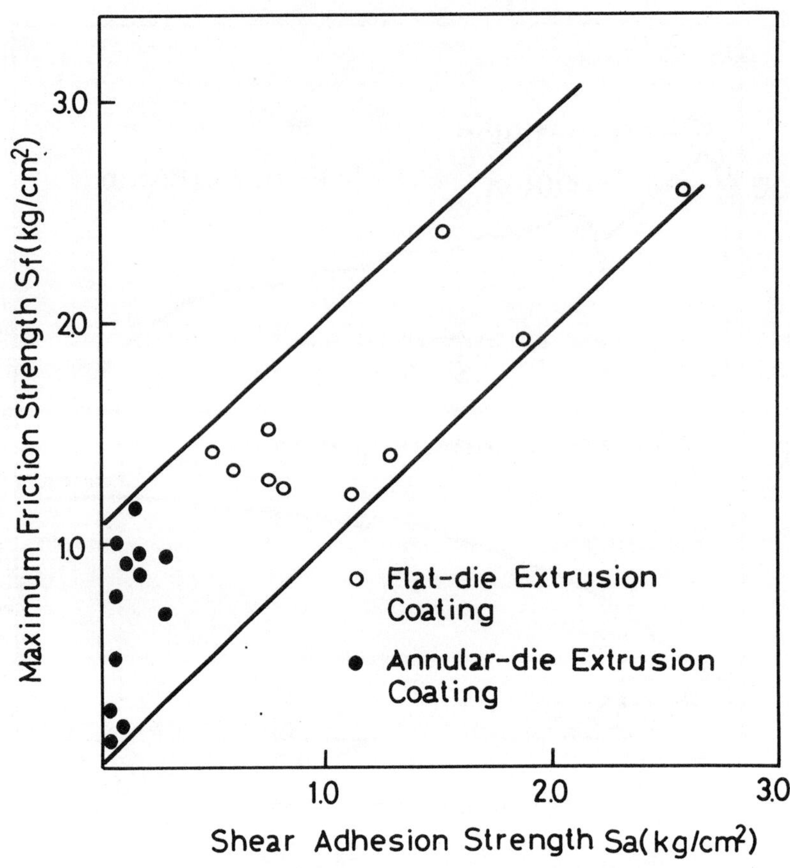

Fig. 11. Relations of displacement ratio of pipes to concrete
jacketings and shear adhesion strength.

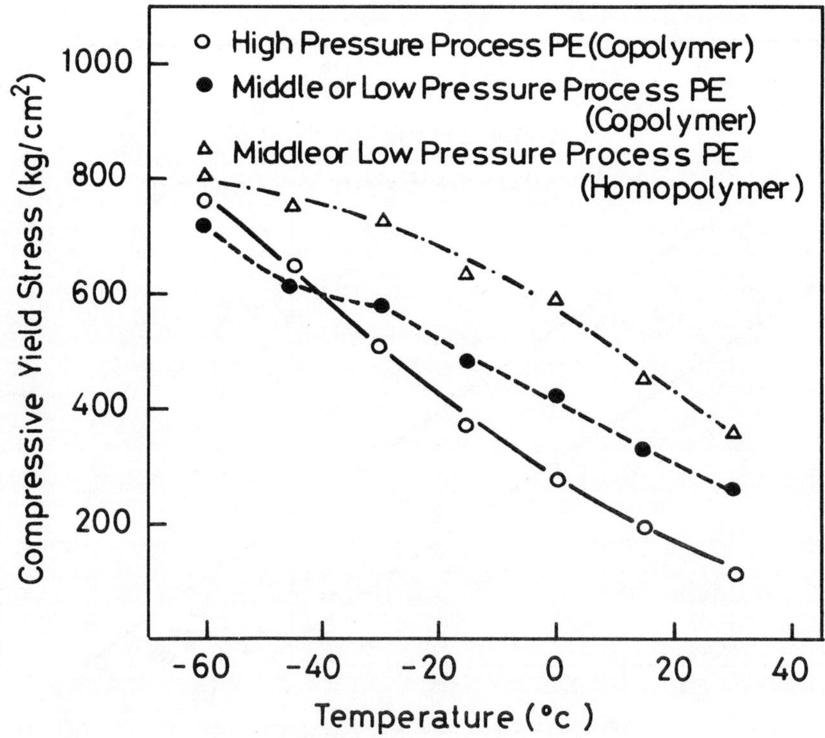

Fig. 12. Relations of shear adhesion strength to maximum
frictional strength.

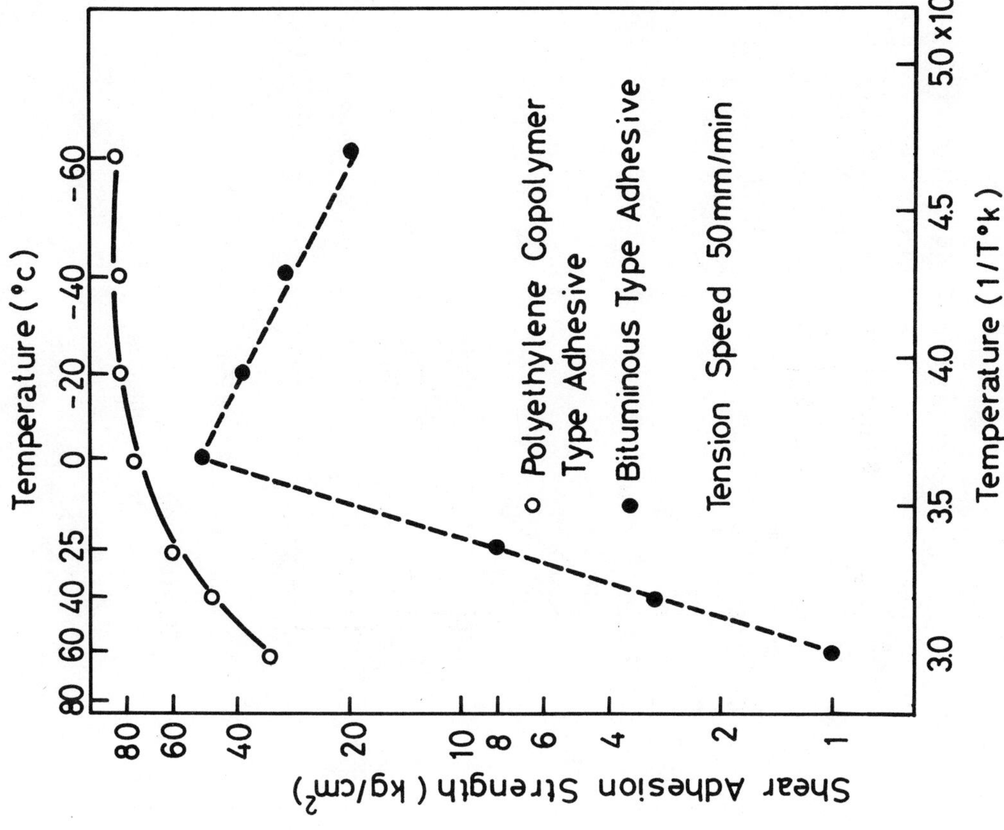

Fig. 14. Temperature dependencies of compressive elastic modulus of various polyethylenes.

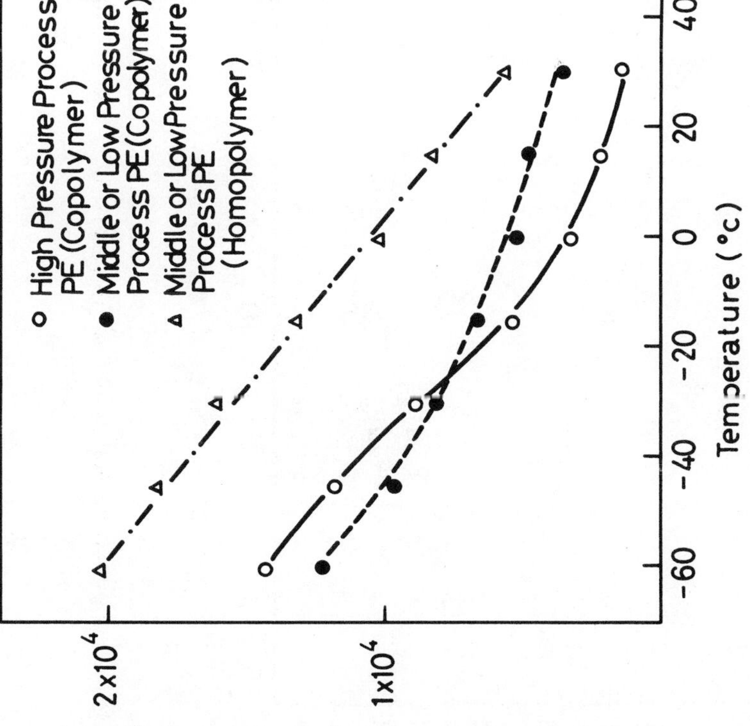

Fig. 13. Temperature dependencies of compressive yield stress of various polyethylenes.

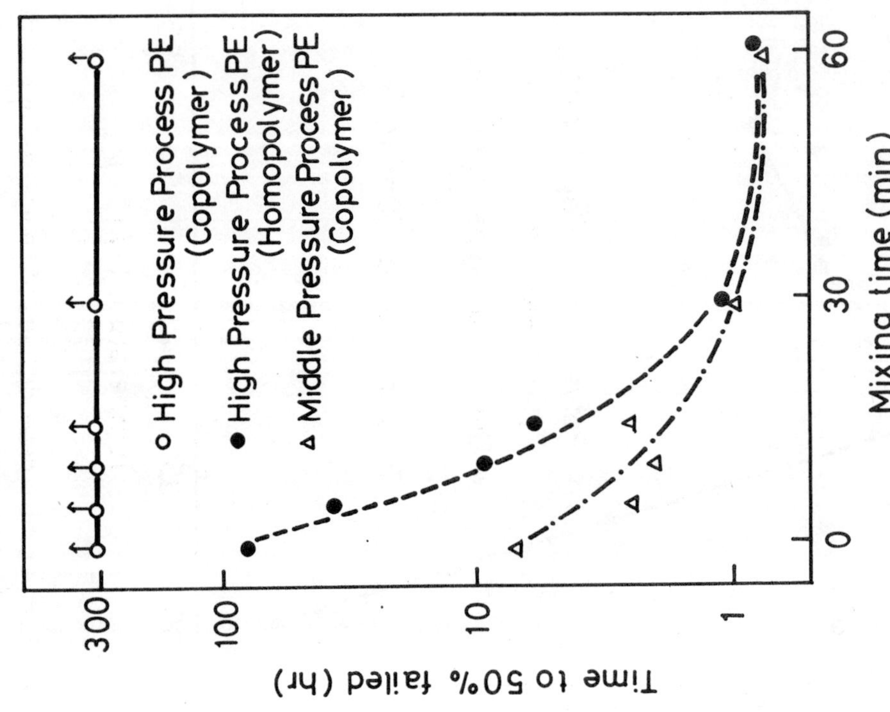

Fig.16. Tensile speed dependencies of shear adhesion strength of two kinds of adhesive.

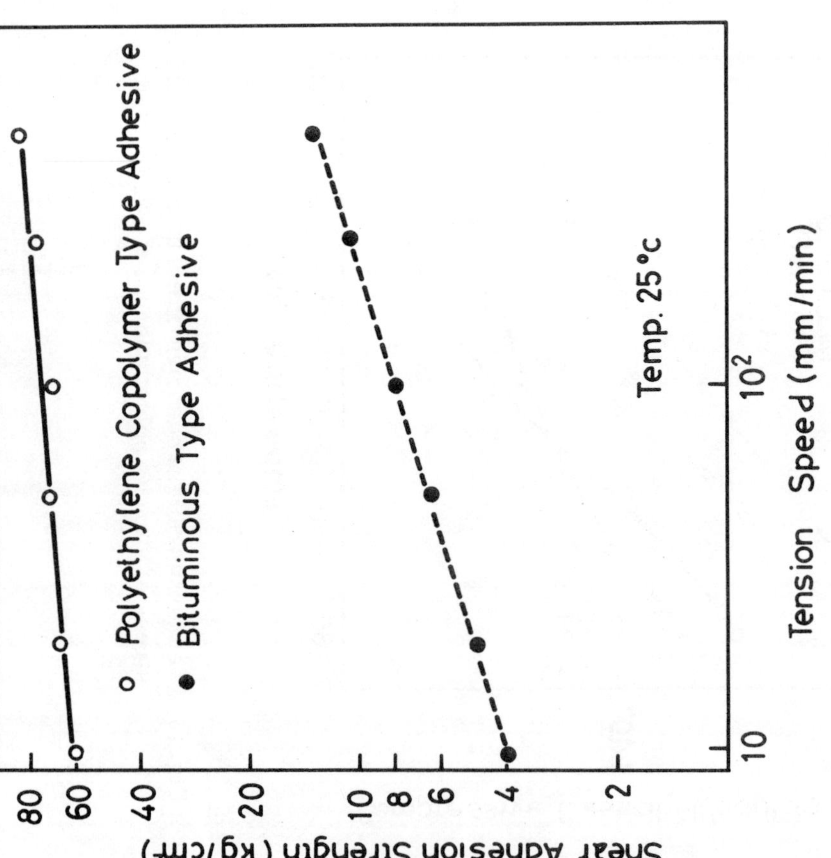

Fig.15. Temperature dependencies of shear adhesion strength of two kinds of adhesive.

internal and external
protection of pipes

September 5th - 7th, 1979

GLASS PIPING SYSTEMS AND THEIR APPLICATIONS

P.W. Bone, B.Sc.

Corning Process Systems, U.K.

Summary

The paper describes the advantages and disadvantages of glass piping systems, the properties of the raw material and the influence they have on design. Alternative types of joint systems are described. Glass is not more widely used as a piping material because it is perceived as being weak and having a limited pressure range. The classical solution has been the use of glass-lined pipe. Recent developments have provided alternatives and these are described. Finally a cost comparison of glass with other corrosion resistant piping systems is given.

Held at Imperial College, London, England.
Organised and Sponsored by BHRA Fluid Engineering

Introduction

The history of man-made glasses dates back to about 4000 B.C.
when it was used as a glaze. Hollow vessels were made by
1500 B.C. probably by moulding. By the first century B.C.
blowing of glass was introduced and this remained the major
method of glass forming for nearly 2,000 years.

Today craft skills still play an important part in the glass
industry. Although the industry can trace its origins back
6000 years it is by no means backward and major glass producers
are still exploring new uses for a fascinating and versatile
raw material. Its uses touch on nearly all aspects of our
domestic and industrial lives.

One specialised use of glass is as a material of construction
in the chemical industry. Complete chemical process plants can
be constructed in glass, and essential to the operation of any
plant are the complex piping systems for the transfer of fluids
between processing units.

Borosilicate Glass - The raw material

A particular glass designated borosilicate glass 3.3 is used
almost exclusively for the construction of glass plant, pipeline and fittings. As with all materials of construction, the
basic properties of the raw material have an influence on the
design of the final product. The raw material, as distinct
from the component made from it, must however always satisfy
certain specified requirements. Glass is no exception. Borosilicate glasses are chosen as raw material because of their
lower coefficients of expansion, which permit the glasses to be
used at higher temperatures and to resist more severe thermal
shock conditions and also because of their greater resistance to
the corrosive effects of acids. Although many different compo-
sitions of borosilicate glass have been developed to meet
special requirements the properties of borosilicate glass 3.3
are controlled by the International Standard ISO 3585-1976. This
standard defines the major properties that are of interest to
users of pipeline; Chemical Resistance, Physical Properties such
as linear thermal coefficient of expansion, density, specific
heat etc and Mechanical Properties such as ultimate tensile
strength, modulus of elasticity and Poisson's ratio.

Glass Pipeline - Its advantages and disadvantages

Glass has now been used as a pipeline material for more than
forty years. The pipeline diameters are standarised at
15,25,40,50,80,100,150mm. The advantages of glass are

- Corrosion resistance - Borosilicate glass is inert to
  almost all materials with the exception of hydrofluoric
  acid, phosphoric acid and hot strong caustic solutions.
  Of these, hydrofluoric acid has the most serious effect,
  even when a solution contains only a few parts per
  million, attack will occur. Phosphoric acid and caustic
  solutions cause no problems when cold but at elevated
  temperatures corrosion occurs.

- It will withstand moderately high operating temperatures.
  Provided borosilicate glass is not subject to rapid
  change in temperature creating undue thermal shock it
  can be used at temperatures up to 300°c. In normal use
  there is a lower limit of 200°c imposed due to the
  limitations of the compatible gasket material P.T.F.E.

- It is noncontaminating, can be sterilised and easily
  cleaned and maintained in a sanitary condition.

- Transparency - operations can be observed.

- In special applications other properties such as UV
  transmission, good electrical properties or precision
  bore become important.

This combination of properties cannot be matched by any other
piping material. But glass has a disadvantage which restricts
its more widespread use because of the safety risk that is
perceived. Glass is a brittle material; it fails abruptly without yield or permament deformation. The spread of breaking
stress is much greater than for metals with the result that the
breaking stress for any one sample cannot be predicted with
accuracy. Failure always results from a tensile component of
stress even when the load is applied in compression. It is now
generally accepted that fractures originate in small imperfections at the surface. A bruise or accidental knock with any hard
object will produce very small cracks on the surface. This will
cause high stress concentration. In metals with ductility,
yield at such a point will tend to equalize the stresses before
failure occurs. There is no relief of stress with glass which
does not yield and fracture can result from the propagation of a
flaw.

The result is that we must accept large safety factors in the
design of glass piping systems. The nominal breaking strength
is approximately 650 bars and 80 bar is taken as the design
stress. This design stress includes residual manufacturing
stresses, and those arising from pressure, temperature and the
coupling.   Operating pressures have been included in
the International Standard which covers glass pipeline compat-
ibility and interchangeability ISO 3587. They are given
together with maximum and minimum wall thickness in table I.

NON DIA	PIPE WITHOUT VALVES	PIPE WITH VALVES	WALL THICKNESS	
			MIN	MAX
MM	bar g	bar g	mm	mm
15	4	3	2.2	3.8
25	4	3	3	5
40	4	3	3	6
50	4	2	3	6
80	3	1.5	3.5	7
100	2	1	4	7
150	2	1	4	8

Coupling Systems

Couplings are a weakness of all piping systems, but they are a
necessary part and ideally must be designed so as not to impose
an operating restriction on the pipeline. Three principle types
of joint have been developed for glass pipeline they are

- The taper joint
- The beaded joint
- The shoulder joint

These joints are illustrated in figures 1,2,3. Each system has
its advantage in terms of cost, ease of use and range of
application. The taper joint was the first system to be used
and is still the most widely used. It is easy to fabricate,
it consists of a conical end moulded or tooled on the glass and
a cast iron slip flange with an isolating compressed asbestos
insert between the flange and the glass. The major disadvantage
of the joint is that due to the plasticity of the insert, there
is a tendency for the joint to relax. This can be satisfactorily
overcome by spring loading the joint.

The beaded joint uses a one piece gaitor coupling. Its limitation
is in higher temperature services due to the material of
construction of the gaitor. It has the advantage that joints can
be site fabricated. Glass naturally forms the rounded bead shape
when rotated in a gas flame, so a piece may be cut to length at
the point of installation and beaded using a simple kit with the
required gauge.

Shoulder joints were developed to overcome the relaxation of taper
joints. The joint itself is more complex and fabrication of the
glass shape is more difficult and expensive.

A number of methods exist to join glass pipeline to other
materials. Direct connections can be made when the mating flange
is true with little or no radius on the inner edge and the bore
matches or is slightly smaller than that of the glass. Adaptor
flanges on the glass can be drilled to suit the mating flange.
When connecting to lined materials direct connection does not
give the glass pipe a full-face joint. In these cases a spacer
of the same bore as the glass pipe fitting within the bolt circle
of the joint is used. Bellows can also be used to make a
misaligned connection or where it is necessary to isolate the glass
pipe from vibration.

Further Developments

Glass is not used more widely as a piping material because it is
perceived as a weak material with a limited pressure range. The
classical solution to this problem has been the use of glass
lined steel. Low carbon steel is used as the base of the pipe,
several layers of glass are then applied in the form of a
vitreous enamel to the cleaned and sand-blasted steel surface
and successively fired. The average thickness of the vitreous
coating is approximately 1.0mm. The expansion coefficient of
glass is very much less than that of steel and the residual
stresses in the glass coating are high. The stresses are compressive and increase the resistance of the coating to cracking
in service. The two major problems encountered with glass-lined
steel are fracture of the lining due to an external impact and

pin-holing from manufacture, both can lead to an undetected
corrosion of the steel.

A development of vitreous enamel lining is to use a more substantial glass inner pipe and shroud it in a steel pipe to
create a pipe-in-pipe system. Here we have the advantage that
the glass pipe being of greater wall thickness is less
susceptable to breakage. Unlike the vitreous enamels, glass
tubing is used to form the inner pipe. It is inserted into
the metal pipe and expanded to form a seal with the metal.
The chemical durability depends upon that of the glass but in
general it is not as good as borosilicate glass.

The latest developments turn away from steel as the strong
component. To use glass itself in the outer pipe. Glass in
the form of glass reinforced plastic. Two systems are available
to cater cost effectively for the different needs of the
chemical industry. The lower cost system operates at the same
pressure as plain glass piping. It has been designed to contain
the process fluids in the event breakage of the pipeline from
an external impact. The system is completely compatible with
plain glass pipe using the same couplings and gaskets and so can
be used as replacements or in conjunction with plain glass. The
system is not designed to add strength to the glass, but to
reduce the dangerous effects of a breakage. It consists of a
thin polyester impregnated glass fibre wrapping applied to
standard plain glass items.

The second system uses a more substantial fibreglass outer-pipe.
This pipe system is designed to operate at higher pressures up
to 10 bar. In this case the system is not interchangeable with
plain glass and a different coupling system is used providing
a seal between both the outer and inner pipes. This type of
pipe is most comparable to the glass-lined steels and plastic
lined pipes. The glass inner pipe provides excellent corrosion
resistance, the fibreglass outer-pipe protects the glass from
impact allowing it to operate safely at higher pressures.

<u>Glass - an economic piping system ?</u>

Engineers specifying piping systems now have a bewildering
choice of materials. The materials vary in performance, initial
installed cost and long term maintenance costs. The installed
cost will depend on;

- Piping material costs

- The size of the line

- The pressure rating or line schedule

- The type of joint used

- The fabrication and installation techniques
  employed and the corresponding labour costs.

- The site conditions and the complexity of the installation

As the importance of these factors varies from case to case any
comparison is at best a generalisation.

Of particular importance on chemical plant is the complexity
of the system and a comparison based on straight runs of pipe
may be of little value in the real world of short runs with
numerous branches and fittings. However some comparison must
be made if the engineer is to meet the challenge of effective
design at an economic price.

Glass is a specialised material for piping systems, its principle
value is as a corrosion resistant or high purity piping system
I have not yet met the engineer who would replace stainless steel
with glass unless he had to. Glass competes with the corrosion
resistant plastic lined pipes and the more exotic materials
such as titanium, zirconium and tantalum.

I will therefore compare plain glass with its competitors PTFE
lined steel Polypropylene lined steel, glass lined steel and
glass fibre reinforced plastic. I include the different members
of the glass piping family wrapped pipe and protected pipe.
Plain glass is taken as one and other material quoted as a cost
ratio.

	Cost Ratio
Plain Glass	1
Wrapped Glass	1.3
Protected Glass	5
GRP Pipe	1.3
Glass lined steel	3.9
Polypropylene lined steel	2
PTFE lined steel	4.2

The cost ratio includes the materials cost for a two inch
system. The cost of flanges nuts, bolts etc are included.
Installation costs are not compared since they vary so much from
case to case.

The above comparison should however be considered against the
maximum operating pressure of the various systems. At the two
inch size the maximum pressure of plain and wrapped glass would
be 3 bar whereas the PTFE lined system would operate at 10 bar.
In specifying piping systems we must therefore consider whether
the higher rating is a benefit for which the higher cost is
justified. In many cases it won't be.

Conclusion

I have presented glass as a unique material for the construction
of piping systems. Its corrosion resistance is excellent and
its commonly assumed drawback of fragility is known and accepted
by the manufacturers. Recent developments of wrapped and
protected glass piping have reduced the potential safety
problems associated with the material. Economically it is an
attractive alternative for systems operating within its pressure
limitations.

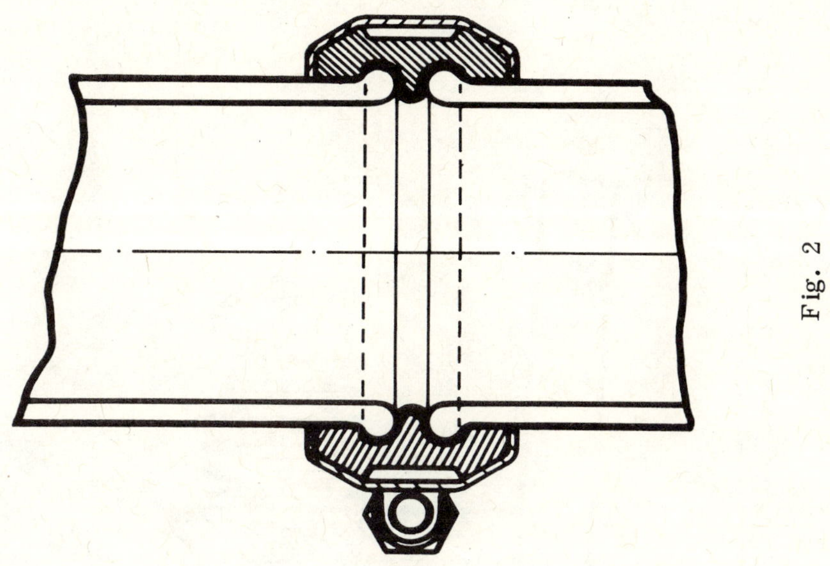

Fig. 2

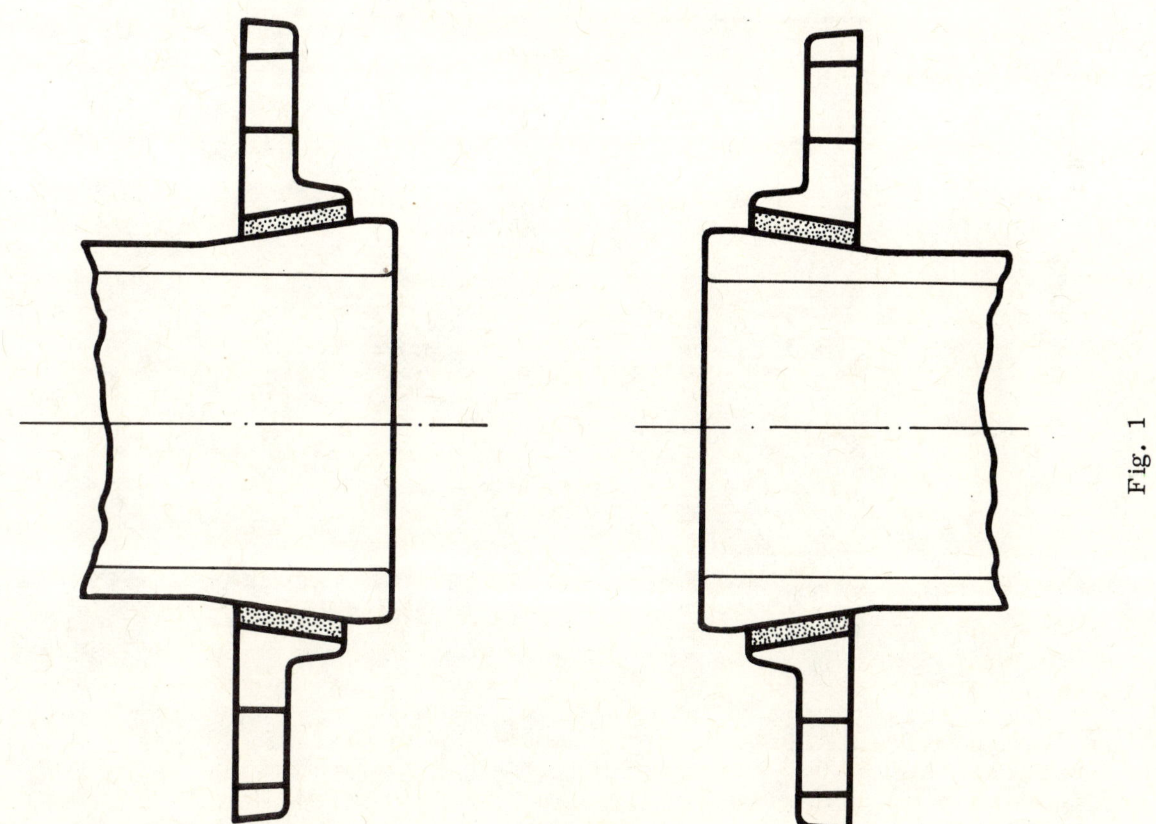

Fig. 1

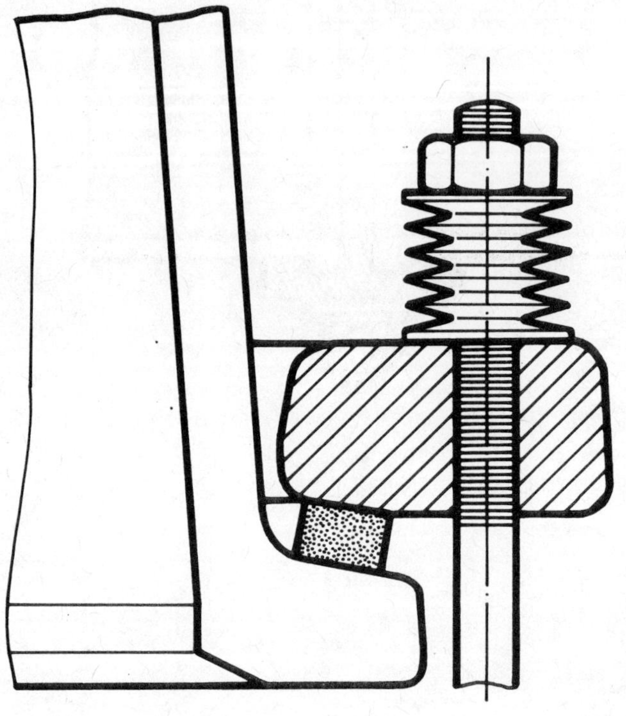

Fig. 3

POLYURETHANE FOR THE CORROSION PROTECTION OF PIPEWORK

J.S. Pitman

Babcock Corrosion Control Ltd, U.K.

Summary

Polyurethanes have been used since the 1960s for the internal protection of pipework against abrasion in the mining and other slurry handling industries, generally under conditions of little or no corrosion and at near-ambient temperatures. Recently introduced polyurethanes exhibit considerably enhanced corrosion resistance, thereby extending their usefulness for pipe linings and coverings where control of corrosion, or corrosion combined with abrasion, is a major requirement.

Held at Imperial College, London, England.
Organised and Sponsored by BHRA Fluid Engineering
Copyright BHRA Fluid Engineering, Cranfield, Bedford, England.

Polyurethanes have been used since the 1960s for the internal protection of pipe-work against abrasion in the mining and slurry handling industries, generally under conditions of little or no corrosion and at near-ambient temperatures. More recently, new polyurethane systems have been developed which exhibit enhanced corrosion resistance, thereby extending their use as pipe linings and coverings where control of corrosion or corrosion combined with abrasion, is a major requirement.

These new polyurethanes are finding increasing applications in process plants where alternative elastomeric materials or other types of lining were previously specified. The wide range of properties available with polyurethanes requires great care in their selection for given applications.

To achieve optimum desired properties the ideal polyurethane would be polyether based to yield a finished product with :

High Tensile Strength at Break	–	45 MpA
Good Elongation at Break	–	500%
Acceptable Tear Strength	–	10 kNm^{-1}
Excellent Abrasion Resistance	–	600 NBS Abrasion Index
Resistance to Chemical Attack	–	Acids, Alkalies, Solvents etc. at elevated temperatures.

In practice all of these properties may not be obtainable with a single polyurethane and therefore it is often necessary to make a compromise.

THE SELECTION PROCESS

The selection process begins by defining the essential properties required for the polyurethane to meet the given duty and then seeking the best available combination of other properties. An indication of the interaction between the various properties is shown as follows :

Property	Interaction between Other Properties
Chemical Resistance	Maximum chemical resistance often associated with lower abrasion resistance and tear strength.
Abrasion Resistance	Best level of abrasion resistance often coincides with maximum tensile strength.
Tear Strength	Increase in hardness normally gives increase in tear strength.
High Temperature Resistance	Maximum temperature resistance required for high speed large particle impact due to energy absorption of the polyurethane surface.
Coefficient of Friction	Low coefficient of friction normally gives best abrasion resistant properties and resistance to cutting.
Tensile Strength	High tensile strength normally gives high abrasion resistance dependent on elasticity.
Elongation at Break	Very high elasticity often reduces abrasion resistance when operating in dry conditions.
Hardness	Very high hardness normally gives high tear strength and reduced abrasion resistance, tensile strength and elongation. Low hardness usually gives reduced abrasion resistance, tensile strength and tear strength.

<u>TEST METHODS AND RESULTS</u>

Testing of polyurethanes is carried out in the laboratory and/or under field service conditions. In order to obtain accurate definitive data, it is essential that samples are correctly prepared. Most polyurethanes are manufactured from two or three component systems and variations in component ratios due to hand weighing or inadequate hand mixing can yield large variations in ultimate properties. Therefore, to obtain satisfactory samples, it is essential that they are prepared using automatic mix dispenser machines to yield consistent products.

<u>CHEMICAL RESISTANCE</u>

It is essential that meticulous testing is carried out over a period of at least one year. Test solutions should be changed regularly and specimen volumes and weights measured at fixed time intervals, preferably every seven days. Behaviour is often temperature dependent and, especially at elevated temperatures, can vary markedly with 10°C increments. Therefore, test temperatures must be carefully controlled to a maximum of plus or minus 1°C.

With elastomers other than polyurethanes it is possible, with experience, to extrapolate test results safely from short term tests (3 months exposure). Most elastomers when vulcanised, consist of a network of long chain molecules distributed in random orientation, cross linked at specific reactive sites. When they undergo swelling by absorption, the consequent internal stress forces re-orientate the random network and the polymer will expand up to the limiting cross link elasticity limit. Normally this mechanism can allow for increases in volume up to as much as 400% provided that the cross links are not chemically attacked. However, for most practical uses it is usually considered that a volume increase of not greater than 15% to reach equilibrium is desirable.

Polyurethanes, being basically block polymers of relatively regular structure, behave in a totally different manner. Due to the regular molecular structure, swelling is limited by the elastic properties of individual bonds. Observable swelling is normally limited to a maximum of 200%. Internal stress states can be reached where further exposure or an increase in temperature can result in catastrophic disintegration. This effect has been found to occur after volume changes of as low as 3.5% dependent on the chemical environment.

The resistance of ester based polyurethanes to hydrolysis is of a much lower order than that of the ether based polyurethanes. Ester based polyurethanes are susceptible to degradation by water due to the relative ease of hydrolysis of the ester group. This group is absent in ether based polyurethanes where the rate of degradation depends on the susceptibility to hydrolysis of the urethane, urea and biuret groups present. Hydrolysis of these groups is much slower and leads to a decrease in molecular weight and cross linking. Ref. 1.

It is postulated that, when disintegration at low volume increase takes place, the loss of cross links and reduction in molecular weight decreases the resistance to permeation. The rate of osmosis through the bulk of the polyurethane increases allowing further hydrolytic attack without yielding significant increase in volume. Ultimately, the whole polyurethane section is affected and suddenly disintegrates, often in a few days. Therefore, extrapolation of short term results to forecast long term behaviour can lead to totally incorrect conclusions.

Examples of the differences in behaviour between individual polyurethanes and other elastomeric polymers are shown in Fig. 1 and Fig. 2.

Fig. 1 illustrates the performance of two polyurethanes compared with a chemical resistant soft natural rubber compound in distilled water at 60°C. In this case the soft natural rubber compound and polyurethane No. 1 have satisfactory resistance reaching an equilibrium state after 400 days with an acceptable volume increase level.

Polyurethane No. 2, shows the behaviour pattern whereby, even though the total volume increase was not excessive, disintegration occurred after 375 days exposure.

Fig. 2 indicates the effect of varying temperature levels on the performance of polyurethane No. 1. Here, performance at temperatures up to $60^{\circ}C$, in a distilled water environment, was satisfactory but at $80^{\circ}C$ breakdown occurred after 290 days immersion.

A basic guide to the performance of those polyurethanes which combine both chemical and abrasion resistance is shown in Fig. 3.

ABRASION RESISTANCE

Standard laboratory tests can only be used on a comparative basis to discriminate between unsuitable materials and promising materials requiring further testing. These laboratory tests are usually carried out on standard equipment such as the National Bureau of Standards Abrasion Tester, the Taber Abrasion Tester or the DIN Abrasion Tester. The equipment is usually designed for dry abrasion tests, although the Taber machine may be used for wet abrasion testing.

Dry abrasion tests are not an infallible guide for the classification of materials for use in wet slurry conditions. The wear rate in abrasive conditions is distinctly affected by heat build up due to the coefficient of friction between the polyurethane surface and the abradent. Coefficients of friction vary between dry and wet conditions leading to changes in the relative performance of various materials.

A classification for dry abrasion tests on typical polyurethanes and soft natural rubbers is shown in Fig. 4.

Further tests are required to assess likely field performance. These can range from basic tests such as rotating samples in a slurry contained in a small test tank up to sophisticated loop test circuits. Ideally it is best to expose a sample in situ under field working conditions.

The particulate nature of the abradent is significant with respect to wear rate. Low mass, uniform particles in a slurry give reasonably regular rates of wear from which fairly accurate life expectancy can be derived.

Relative abrasive wear characteristics of different materials, including pipe linings, are determined as a function of slurry velocity. Tests have been carried out using iron ore of mean particle size 150mm., maximum 1.5mm., as the abrasive solids in suspension. In terms of mean slurry velocity and average concentration of solids by volume it was shown that different polyurethanes of approximately equivalent NBS abrasive index have close relative wear rates.

Conclusions were drawn giving a ranking of materials tested in order of wear rate compared against a mild steel reference. Data indicated that standard polyurethanes had an expected order of life approximately thirteen times that of mild steel under equivalent working conditions. Ref. 2.

Large mass particles of irregular shape can cause deep cutting of the polyurethane surface leading to a much faster rate of material loss. Another factor involved is that of flash impact temperature. When a particle of relatively large mass moving at speed strikes a surface, some of its energy will be converted into heat. As polyurethanes have generally poor hysteresis properties, dissipation of heat is slow. On impact, by a large particle, the local temperature can rise, if only for a fraction of a second, to over $100^{\circ}C$. If the polyurethane has poor heat resistance, surface degradation can occur leading to lowering of abrasion resistance.

An examination of other physical properties in relation to abrasion resistance indicates that certain correlations can be drawn. In most polyether based polyurethanes those with best abrasion resistance have hardnesses between 75° and 95°

Shore 'A'. Outside this hardness range abrasion resistance tends to decrease. Relationships also exist between abrasion resistance, tensile strength, elongation and tear strengths. The results obtained on a typical family of polyurethanes are shown in Fig. 5 and Fig. 6.

When operating in an aggressive chemical environment, the effects of absorption or chemical attack can radically change abrasive wear resistance. If the exposed surface is heavily absorbed or a chemical change in structure takes place, then the stress/strain properties and tear resistance will alter, normally increasing wear rate. Therefore, when assessing abrasion resistance for given corrosive environments, it is necessary to consider the effects of the chemical system on the polyurethane.

CHEMICAL AND ABRASION RESISTANT POLYURETHANE PIPE LININGS

Polyurethane cast elastomers are ideally suited to the centrifugal method of lining straight pipes. By such methods it is possible to produce linings up to 12mm. thick.

With respect to pipe fittings, bends, tees, crosses etc., the production method is either to mould the lining using internal formers or apply the polyurethane by a spray application. The moulding method is generally the best technique but has the disadvantage of requiring somewhat expensive formers especially if non standard geometry fittings are required.

Spray application of polyurethanes enables non standard geometry fittings etc. to be lined or covered. It is, however, essential that the correct physical properties are imparted to spray applied coatings if they are to be used for abrasion resistance/severe corrosion conditions. This requirement often renders solvent based systems unsuitable for demanding duties due to retention of a proportion of the solvent.

However applied, the success of bonded polyurethane coatings is frequently as dependent on the quality of the substrate preparation as it is on the polyurethane used.

There are many service conditions in industry where abrasion causes severe wear to process pipework. Such abrasive wear can occur rapidly, often within a few weeks and therefore leads to high costs for maintenance and replacement of equipment. Aggressive chemical environments compound these problems as failure of equipment can lead to costly contamination of the surrounding plant environment.

Being sacrificial in their mode of operation, abrasion resistant protective linings involve a consideration of both initial costs, and of cost per year of life. Besides the cost of the component or lining, increased life results in lower mainten-ance and failure downtime, thereby conferring cost advantages to the plant operator.

The relative costs and life expectancy for different types of pipe protection systems are shown in Fig. 7.

The life/cost ratio, although simple, is not a comprehensive measure of the advantage of one system over another. For example, it takes account of capital cost only, neglecting the cost of installation and maintenance downtime. As a general rule, the more severe the abrasion, the greater the relative advantage of more durable linings.

Similar considerations apply to polyurethanes used solely in a corrosion resistant role. In many instances, especially with 100% solids polyurethanes, robustness, ease of maintenance and repair, and extended lining life, more than justify the initial material costs, to give an overall cost effective protective system. On this basis, polyurethane competes favourably with other protective systems used in comparable corrosive environments.

<u>INDUSTRIAL APPLICATIONS</u>

Polyurethanes have a wide potential use as protective pipe linings in industry where their excellent resistance to abrasion coupled with their chemical resistance, offer advantages to the user. Examples of use include, the transportation and treatment of water, sea water cooling mains, sewage and industrial effluent, solvent extraction processes in hydrometallurgy, pulp and paper processing, mining, ash and tailings disposal and general transportation of solids.

Some examples of the use of polyurethanes for the protection of pipes are cited below :

(a) Transportation of 40% solids slurry waste to disposal from molybdenum process plant. After six years linings in excellent condition with minimal wear.

(b) Hydrometallurgical copper extraction pipework handling copper sulphate, crushed solids and sulphuric acid. To date, five years satisfactory service.

(c) Pipework carrying calcium sulphate slurry at temperatures up to 70°C. Pipes lined with abrasion resistant polyurethane, outside coated with corrosion resistant polyurethane. Pipes have operated successfully to date whereas unprotected mild steel pipe failed in three to six months.

(d) Sand slurry and fine gravel pipelines working satisfactorily after three years service compared to three months life for unlined mild steel.

(e) Chemical resistant polyurethane lined pipework handling environments as follows :

Sodium chloride solution at 60°C. Four years satisfactory life to date.

Acid/alkaline effluent plus oil at 40°C. Two years satisfactory life to date.

12% hydrochloric acid, 50% phosphoric acid mixture at 20°C. Four years satisfactory life to date.

Uranyl sulphate plus 10% sulphuric acid and kerosene at 45-60°C. Two years satisfactory life to date.

20% sulphuric acid at 60°C. Five years satisfactory life to date.

<u>CONCLUSIONS</u>

Modern polyurethanes are able to combine excellent corrosion and abrasion resistant properties in a single material. As with all protective systems, it is essential to match the properties of the selected polyurethane to the process conditions to be encountered. Clearly, this matching can only be carried out if comprehensive data is available from laboratory and field tests, and also if the behaviour of the selected material under different conditions is well understood.

Providing that the correct selection process is followed, then polyurethanes are able to offer operators of process plant equipment both technical and economic advantages over conventional materials.

<u>REFERENCES</u>

1. Robert J. Athey. "Water Resistance of Liquid Urethane Vulcanisates."
        Du Pont Bulletin No. 5, April 1965.

2. J. Boothroyde and B.E.A. Jacobs. "Pipe Wear Testing, 1976-1977"
        B.H.R.A. PR 1448, December 1977.

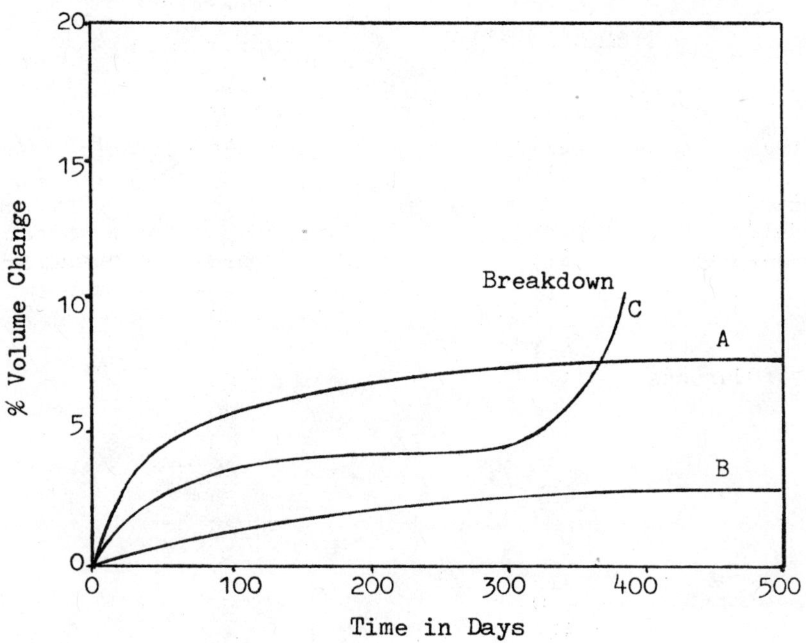

Fig. 1.   Distilled Water at 60°C.
A. Soft Natural Rubber
B. Polyurethane No. 1
C. Polyurethane No. 2

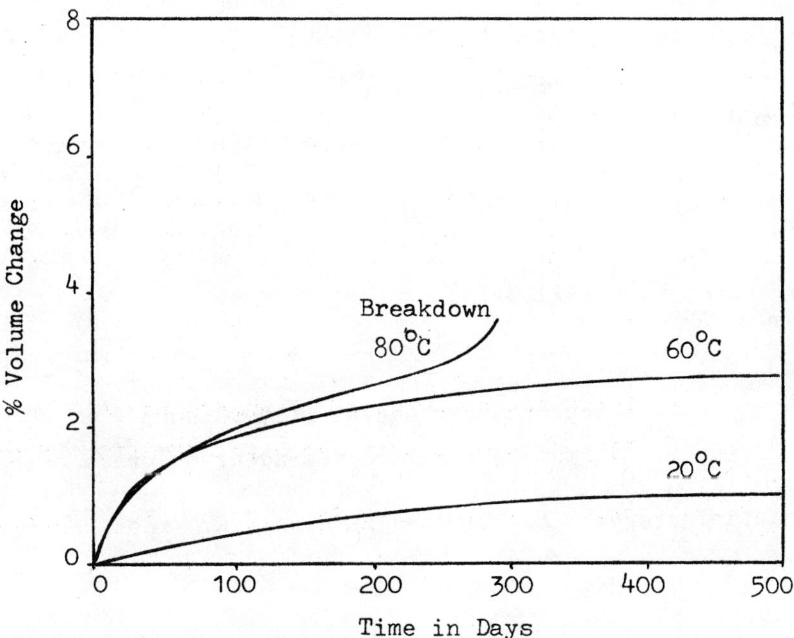

Fig. 2.   Polyurethane No. 1 in Distilled Water
at various temperatures.

FIGURE 3. CHEMICAL RESISTANCE. TABLE OF PERFORMANCE

Chemical	% Concentration	Temperature in °C.	Resistance Rating
Acetic Acid	20	20	L
Alcohols	All	20	L
Aluminium Salts	All	80	S(2)
Alums	All	80	S(2)
Ammonium Chloride	25	20	S
Ammonium Hydroxide	10	20	S(1)
Aromatic Hydrocarbons	All	20	N.R.
Barium Salts	All	80	S(2)
Brine	All	80	S(2)
Calcium Salts	All	80	S(2)
Chlorinated Hydrocarbons	All	20	N.R.
Chromic Acid	10	20	L
Copper Salts	All	80	S(2)
Detergents	20	20	S
Esters	All	20	N.R.
Ferric Chloride	20	20	S
Ferric Salts	All	80	S(2)
Ferrous Salts	All	80	S(2)
Formic Acid	20	20	L
Halogens	All	20	L
Hydrobromic Acid	25	20	S
Hydrochloric Acid	50	20	S
Hydrofluoric Acid	10	20	L
Hydrogen Peroxide	10	60	S
Kerosene	All	20	S(1)
Ketones	All	20	N.R.
Magnesium Salts	All	80	S(2)
Mineral Oils	All	20	S
Nickel Salts	All	80	S(2)
Nitric Acid	5	20	L
Phosphoric Acid	70	20	S(1)
Potassium Hydroxide	25	20	S(1)
Potassium Salts	All	80	S(2)
Sodium Hydroxide	25	20	S(1)
Sodium Hypochlorite	All	20	L
Sodium Salts	All	80	S(2)
Sulphuric Acid	30	20	S
Vegetable Oils	All	20	S
Water, Distilled		60	S(2)

NOTES ON TABLE: 1. Dependent on grade of polyurethane

2. Maximum temperature dependent on grade of polyurethane

SYMBOLS: S - Satisfactory,   L - Limited Life,   N.R. - Not Recommended

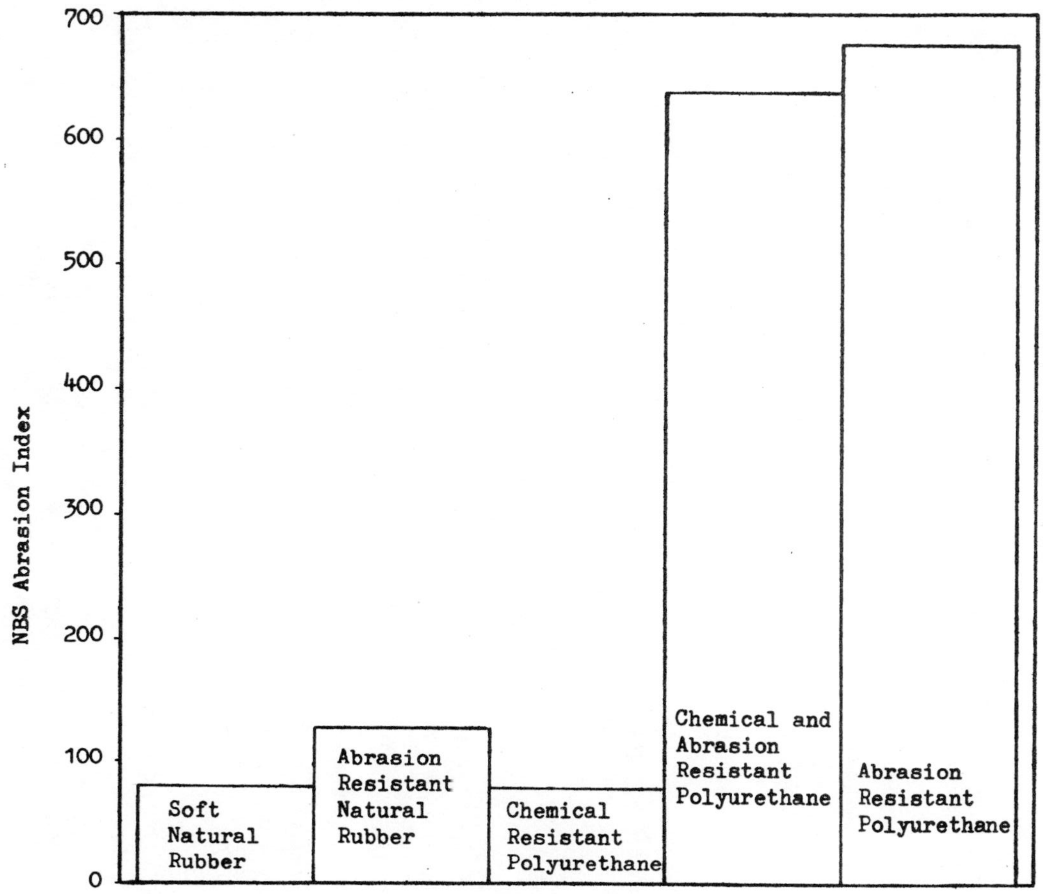

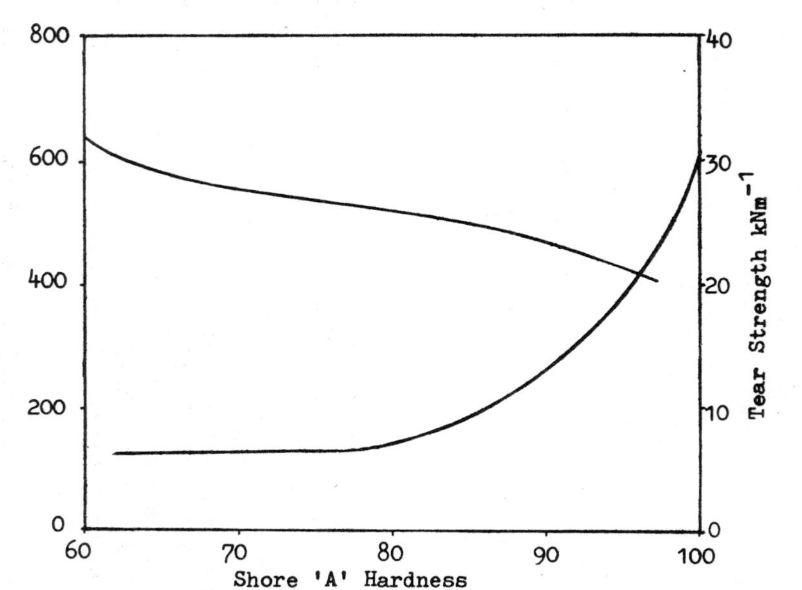

Fig. 4.   Classification for Dry Abrasion Tests on Typical Polyurethanes and Soft Natural Rubbers.

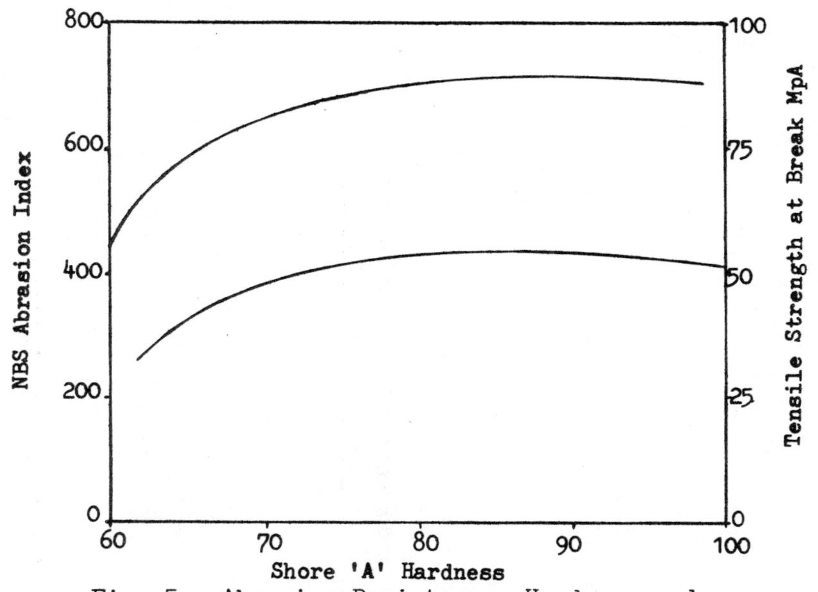

Fig. 5.   Abrasion Resistance, Hardness and Tensile Strength Relationship.

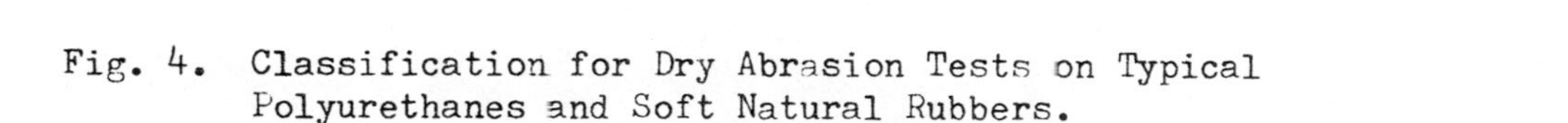

Fig. 6.   Elongation at Break, Hardness and Tear Strength Relationship.

FIG. 7.   RELATIVE COSTS AND LIFE EXPECTANCY

Pipe Material System	Relative Life	Relative Cost	Life/Cost Ratio
12.5mm. Wall Mild Steel	1	1	1
6mm. Wall Mild Steel Lined 6mm. thick Rubber	2.5	1.25	2
6mm. Wall Mild Steel Lined 6mm. thick Polyurethane No. 1	4	1.25	3.2
6mm. Wall Mild Steel Lined 6mm. thick Polyurethane No. 2	8	1.30	6.15

All relations based on 250mm. nominal bore, 6 metres long flanged pipe.

internal and external protection of pipes

September 5th - 7th, 1979

CONCRETE INTERCEPTOR SEWER CORROSION PROTECTION—
A STATE OF THE ART REPORT

K.K. Kienow, P.E., MASCE, Q.C.P.

Hydro Conduit Corporation, U.S.A.

Summary

This paper presents a discussion of the development of design methods for concrete sanitary sewer internal corrosion protection.

Early formulas which served as qualitative guides for sulfide control in sanitary sewer design are discussed as are later studies which attempted to define the mechanisms of sewer corrosion. These studies were pursued when it became apparent that there was more to the problem than just protecting the interior of the pipe from corrosion. "Sewer gas" (hydrogen sulfide) was recognized as a hazard to treatment plants and maintenance personnel, destructive of treatment plant facilities, electrical switch gear, wet wells, manholes and pumps and motors in lift stations.

Qualitative guidelines for sulfide control in sewer systems were published in 1974 by the United States Environmental Protection Agency. Limiting slope/flow relationships are discussed along with methods for calculating concrete corrosion rates due to a particular level of sulfides.

A quantitative sulfide prediction equation was developed in 1977. This equation has been used extensively in the United States to predict sulfide build-up levels. "Life factor" pipe design principles are discussed. This method is used to design a corrosion resistant sewer pipe as an alternate to clay pipe or lined concrete pipe. Twenty major projects have been constructed in the United States involving 140 km of "Life factor" pipe from one half meter to several meters in diameter.

This paper includes a discussion of a case study performed on a severely corroded 22 km interceptor sewer built in 1963. Measured sulfide levels and observed pipe interior corrosion were compared with predicted sulfides and corrosion using the equations now available. The study showed excellent correlation between predicted and observed values. Both the 1977 sulfide prediction method and the 1974 EPA corrosion equation can be used with a good level of confidence. Significant cost savings, as much as 25% of the project cost, have been realized by agencies which have ultilized the new technology.

Held at Imperial College, London, England.
Organised and Sponsored by BHRA Fluid Engineering
Copyright BHRA Fluid Engineering, Cranfield, Bedford, England.

NOMENCLATURE

A	Alkalinity of concrete cover over reinforcing steel.
b	Surface width of stream, m.
BOD	Biochemical oxygen demand mg/l.
EBOD	Effective BOD mg/l.
BOD_c	Climactic BOD mg/l.
$EBOD_c$	Climactic EBOD mg/l.
c	Average corrosion rate mm/yr.
D	Diameter, m
d_m	Mean depth, m.
k	Acid loss coefficient.
L	Design life, years.
M'	Coefficient for Pomeroy-Parkhurst formula – sulfide build-up term
N	Coefficient for Pomeroy-Parkhurst formula – sulfide loss term
P	Wetted perimeter, m.
ϕ_{sw}	Sulfide flux to the pipe wall, gm/m^2-hr.
Q	Flow, litres per second.
r	Hydraulic radius, m.
s	Slope.
S_2	Downstream total sulfide level mg/l.
S_∞	Limiting sulfide level mg/l.
(s)	Sulfide level mg/l
T	Temperature degrees C.
u	Velocity of stream m/sec.
z	Thickness of concrete cover over reinforcing steel in the crown of the pipe, mm.
Z	Pomeroy's Z

<u>HISTORY OF SEWERS</u>

One of the earliest sewers, constructed about 3750 B.C., was unearthed during excavation in Nippur, India. An excavation at Tell Asmar, near Bagdad, disclosed a sewer constructed in 2600 B.C. About 1700 B.C., the Minoans, who lived in Crete, had installed an elaborate system of stone drains which carried roof drainage, sewage and storm waters. The famous Roman drain called the Cloaca Maxima was in use from about 200 B.C. until the early 1900's. Most of these ancient sewers (with the exception of the Minoans system) were primarily for the collection of general drainage water. When waste accumulated in the streets and was washed into the drains with the storm water, the first "combined sewers" went into operation.

Modern sewage began in Europe in the late 19th century as outbreaks of Asiatic cholera decimated the population. In London, sessions of Parliament were disrupted by the foul odors eminating from the Thames River, as smaller sewers collected the wastes of millions of people and discharged them to the River. In the United States, many cities were without sanitary sewers until 1920, and today, we are still at work separating combined sewer systems which discharge waste to our rivers and bays during periods of extensive rainfall.

<u>CONCRETE PIPE SEWERS</u>

Concrete sewers date back to Roman times and in the mid-19th century, large sewers in Paris were constructed of stone masonry with a heavy cement plaster layer on the interior. Most of the concrete sewer pipe installed in the United States during the 1870's and 80's is still in service. In 1962, a 100-year old sanitary sewer in St. Louis was uncovered and examined. The line was found to be in excellent condition. In St. Paul, Minnesota, some 28,700 m (94,000 feet) of 229 mm. to 711 mm. (9 inch to 28 inch) concrete sewers installed between 1875 and 1888 are still in service. Most of the trunk and interceptors in the city's 1420 km. (882 mile) sewer system are concrete pipe.

One hundred twenty million pounds (240 million dollars) worth of interceptor sewers are currently under construction in Sacramento, California. The main City Interceptor's 2.59 m (102") Reinforced Concrete Pipe (RCP) and the 3.05 m (120") outfall line from the new 83 million pound (175 million dollar) Regional Treatment Facility are completed. Concrete pipe is planned in the Sacramento system for the entire 2.59 m to .46 m (102 inch to 18 inch) size range. Major concrete interceptor sewers (without liners) are currently being installed in Dallas, Texas; Las Vegas (Clark County), Nevada; Los Angeles County, California; Tucson, Arizona; Seattle, Washington; and Colorado Springs, Colorado. In 1978, over one million pounds (200 million dollars) in concrete pipe sanitary sewers was installed in the United States. Most of this pipe was unlined conventional ASTM C76 RCP.

<u>SULFIDE DAMAGE TO CONCRETE SEWERS</u>

With a long and continuing history of successful use, why have we occasionally heard of concrete pipe sewer failures after twenty or even ten years of use?

Under certain conditions, hydrogen sulfide (H_2S) is formed in the sewage. As the H_2S concentration in the sewage increases, some of the H_2S escapes to the sewer atmosphere and is deposited on the sewer wall above the flowline. Bacteria oxidize the H_2S to sulfuric acid. The acid is corrosive to many materials, including concrete. Manhole steps disintegrate, treatment plant structures and equipment are damaged, and sludge bulking may occur in the activated sludge process. H_2S gas is lethal in very small concentrations and is made even more deadly by the fact that the first "sniff" destroys the effectiveness of the olfactory system and the ability of the nose to detect the gas ceases. Many sewer maintenance workers have been killed by hydrogen sulfide. Recently in Corona, California, several young boys at play removed a manhole cover and entered the manhole. Several firemen were injured in the attempted rescue, and one of the boys was killed by the lethal gas.

Design methods to avoid production of H_2S in sanitary sewers have been quite well defined for years as have methods of controlling sulfides when their generation could not be avoided. Most failures of concrete pipe sewers designed within the last twenty five years are the result of either poor engineering design, improper installation or lack of adequate sewer maintenance.

DESIGN PRINCIPLES FOR AVOIDING SULFIDE BUILD-UP

The basic design principles for a sulfide-free system have been known for the past thirty-five years or more. Pomeroy and Bowlus (Ref. 1) reported on some of the earlier work in 1946. While the older formulas for avoiding sulfides have been "fine tuned" and improved over the years, it has been reported by the United States Environmental Protection Agency (USEPA) that "serious sulfide problems have been avoided where any one of the earlier formulas was followed in sewer design". (Ref. 2) Force mains, surcharged sewers, or inverted siphons are the principal source of sulfide generation in most systems where significant sulfides are observed. The typical failure occurs in a gravity sewer downstream of a force main discharge as sulfides built up in the force main are released in the gravity line.

Poor grade control on early pipe installations often resulted in "sags" in which sewage ponded, solids built up in the invert, and sulfide production turned a section of the line into the "world's longest septic tank". Modern installation methods, including laser grade control, have eliminated the problem.

Even when extremely flat grades exist, balling the line at yearly intervals has a marked effect in reducing sulfide levels and attendant problems. The USEPA has ruled (Ref. 3) that protecting the pipe by lining or using a more expensive product (vitrified clay pipes) cannot be justified on the basis of saving future sewer maintenance costs. Proper maintenance of the system such as routine line cleaning must be assumed by the designer. The USEPA provides as much as $87\frac{1}{2}\%$ of the cost for initial construction and the local agency pays 100% of the operation and maintenance costs. The temptation to build a "platinum plated" expensive system and "save" local money by "skimping" on routine maintenance is very real.

MECHANISM OF SULFIDE GENERATION

The following discussion of the processes involved in sulfide generation is taken from Ref. 2:

"For sulfate to be reduced to sulfide requires a medium completely devoid of free oxygen or other active oxidizing agent. The stream of wastewater in a partly filled sewer is not completely anaerobic, because it is exposed to the atmosphere. Oxygen absorbed at the surface of the stream generally reacts quite rapidly and in large sewers, its concentration is held to a very low level, yet enough is present to prevent sulfate reduction in the stream. The place where strictly anaerobic conditions can develop is in the slime layer that forms on the submerged pipe wall . . . ." A typical slime layer may be considered to be 1.0 mm (0.04 inch) thick, but, if the velocity of the stream is high, it may be no more than 0.25 mm (0.01 inch) thick, and if abrasive material is carried by the water, the wall may be scoured clean. At low velocities, the slime may be as much as 3.0 mm (1/8 inch) thick, or even more.

If oxygen is present in the system, it diffuses into the slime layer, but the aerobic bacteria found there use it so rapidly that it is depleted in a very short distance. Except where oxygen concentrations are high, the aerobic zone is less than 0.25 mm (0.01 inch) thick. Beneath that, the slime layer is anaerobic, and it is there that sulfide generation occurs.

Calculations based upon diffusion coefficients and observed rates of sulfide production show that sulfate and/or the organic nutrients available to the sulfate-reducing bacteria are used up in a very short distance and that the thickness of the sulfide-producing zone is generally on the order of 0.25 mm (0.01 inch). At deeper levels, the slime layer is anaerobic but largely inactive because of the lack of a nutrient supply.

As long as the face of the slime layer is aerobic, sulfide diffusing out of the anaerobic zone will be oxidized there. No sulfide will be found in the stream unless it is from extraneous or upstream source. Fig. 1 illustrates the processes just described, with the slime layer depicted on a greatly magnified scale.

It is the oxygen supply, rather than the kind of bacteria present, that determines where the aerobic zone ends and the anaerobic zone begins. The oxygen concentration in the stream, as well as temperature and the concentration of organic food materials, determine how deep the oxygen will penetrate. If the oxygen supply increases, the aerobic zone will become thicker.

"If the oxygen concentration in the stream drops to a low level, generally a few tenths of a mg/l, not enough oxygen will reach the surface of the slime layer to oxidize all of the sulfide that is produced. Sulfide can then escape into the stream." The process is illustrated in Fig. 2.

"The oxygen concentration that is critical for preventing sulfide access to the stream is generally in the range of 0.1 to 1.0 mg/l. It depends, among other things, on temperature and on the turbulence of the stream. If the water is stationary, or moving slowly, oxygen becomes depleted near the pipe wall and sulfide may escape from the slime layer even when the main bulk of the wastewater contains several milligrams per liter of oxygen. If there are organic solids slowly rolling along the bottom of the pipe, they will release sulfide even when the stream has a high oxygen content. When the oxygen content is low enough so that sulfide leaks into the stream, it may nevertheless be only a minor part of it that gets through. Completely anaerobic conditions must be approached before all the sulfide produced can pass into the stream."

DESIGNING FOR CONTROL OF SULFIDES

Over the past thirty-three years, many attempts have been made to define parameters which would predict, in a qualitative way, the sulfide build-up potential of a gravity sewer line.

One of the earliest (1946) of these, by Pomeroy and Bowlus (Ref. 1) was limited to pipes flowing not over half full. The method defined a marginal velocity above which sulfide build-up would not be expected. The marginal velocity was proportional to the Effective BOD, BOD x 1.07^{T-20} where T represents the sewage temperature in degrees Celsius. In 1950, Davy (Ref. 4) developed an equation which took into account the relative depth of flow but used a different assumption with regard to the effect of velocity on oxygen absorption. The Davy equation was modified and combined with a hydraulic equation and became known as the Pomeroy "Z" formula. Pomeroy first published the "Z" formula (Ref. 5) in 1970, although it had been in use for some years prior to 1970. Thistlethwayte (Ref. 6) recommends the "Z" formula for preliminary screening purposes. Thistlethwayte presents a complex formula for calculating sulfide build-up. Pomeroy's "Z" formula is as follows:

$$Z = \frac{3\ EBOD}{s^{1/2}\ Q^{2/3}} \times \frac{P}{b} \qquad \text{where}$$

EBOD = effective BOD mg/l
s = slope of the energy line of the stream
Q = wastewater flow, litres/sec.
P = wetted perimeter, m.
b = surface width of stream, m.

Sulfide build-up is common where "Z" exceeds 10,000, but is rare where "Z" is below 5,000.

EFFECT OF VELOCITY ON SULFIDE BUILD-UP

Thistlethwayte (Ref. 6) emphasizes the velocity at which slimes will be sheared from the pipe wall. If sufficient velocity can be maintained, slimes will be sheared from the wall and sulfide build-up will be minimized; however, Pomeroy (Ref. 2) states that:

"Observations of the effect of velocity on the slime layer are few. The slime layer varies in thickness at different locations on the submerged surfaces (Ref. 7) but the minimum thickness for unimpaired sulfide generation is not known. It was shown (Ref. 8) that a velocity of 7 ft/sec did not impair the ability of a slime layer to use oxygen at a maximum rate. It was also found (Ref. 9) that sulfide build-up in a pressure main where the pumping velocity was 4 ft/sec was no less than in mains where the velocity was less. In view of the above work, it would not appear safe to rely upon velocity to keep slime layer thin enough to effectively control sulfide generation. However, the slime layer may be reduced to a few mils, and at that thickness a rather small oxygen concentration might keep it aerobic or at least would prevent the small amount of sulfide that might be formed from getting into the stream.

The effect of velocity on the transport of solids is more tangible. A small amount of gritty matter lying in the pipe does not have much effect on sulfide generation, but when the velocity is slower, organic solids may concentrate at the bottom, intermittently sliding and rolling along. Because of oxygen depletion in the loosely deposited solids, and the large interfacial area, sulfide generation is then markedly accelerated. If the velocity is slow enough so that a deep, stable deposit of sludge forms, it does not proportionally increase sulfide generation, because the sludge becomes starved for sulfate, but conditions do continue to become worse with decreasing velocity."

Fig. 3 illustrates the effects of varying velocity.

EPA FLOW-SLOPE CRITERIA

The most recent qualitative guideline for the design of sulfide-free gravity sewer systems was published by the USEPA in 1974. The "Process Design Manual for Sulfide Control in Sanitary Sewerage Systems" (Ref. 2) was prepared by Dr. Pomeroy under a contract with USEPA. The Manual presents a guide (Fig. 4) titled, "Flow-Slope Relationships as Guides to Sulfide Forecasting". Fig. 4 is limited to sewers with flow depths of two thirds pipe diameter or less. There cannot be any force mains operating unless measures are taken to reduce their sulfide producing and oxygen depleting potential.

The EPA Manual gives the interpretation of Fig. 4 as follows:

"The interpretation of Fig. 4 must be qualified as being related to wastewater streams of specified characteristics. The temperature of the wastewater in a sewer follows an annual cycle, BOD and quantity of flow have diurnal cycles, and flows usually have long-time trends. It is useful to define a climactic condition as the combination of the average temperature for the warmest three months of the year. Where diurnal BOD curves have not been made, it may be assumed that the BOD for the six hour high-flow period is 1.25 times the BOD of a flow-proportioned 24 hour composite. The climactic EBOD (effective biochemical oxygen demand) is defined by the equation:

$$(EBOD)_c = (BOD)_c \times 1.07^{(T_c - 20)}$$

where:

$(EBOD)_c$ = climactic EBOD, mg/1
$(BOD)_c$ = climactic BOD, mg/1
T_c = climactic temperature, deg. C

Let it be assumed that the climactic EBOD for a sewer is 500 mg/l. Then the interpretation of Fig. 4 is as follows:

Curve A: While the climactic condition prevails, a system functioning with slope-flow
relationships as shown by Curve A may be expected to produce very little sulfide,
rarely more than 0.1 or 0.2 mg/l of dissolved sulfide. The annual average dissolved
sulfide concentration is expected to be only a few hundredths of a mg/l.

Curve B: While the climactic condition prevails, a system functioning with slope-
flow relationships as shown by Curve B may produce dissolved sulfide at concentrations
of several tenths of a mg/l.

It must be emphasized that Fig. 4 cannot be used to accurately predict sulfide conditions. The purpose is to indicate an apparent trend in the effect of flow-slope relationships on sulfide build-up. Additional data are necessary to refine and fully
substantiate the indicated relationships.

If the climactic EBOD is higher or lower than 500, the positions of the curves will
be altered, except that the cut-off effective slope of 0.6 percent for Curve A at
small flows will remain the same, since the determining factors are hydraulic. The
relationship between EBOD and required slope is complex. The nutrients available for
the sulfide generating bacteria in the slime layer are presumed to be roughly proportional to the BOD, but the oxygen requirement to prevent sulfide build-up may not increase in the same ratio. If slope, and hence velocity, is increased, the greater
turbulence not only increases oxygen absorption, but also provides more effective
transfer of oxygen to the slime surface. As a result, a lower oxygen concentration
suffices to prevent the escape of sulfide from the slime layer into the stream.
Equally important, the length of time that the wastewater spends in transit is
reduced.

Considering all factors, it is judged that similar sulfide conditions will result if
the effective slopes are increased or decreased in proportion of the square root of
the effective BOD. Thus, Fig. 4 shows that for a Curve A condition, a flow of 2.0 cfs
requires an effective slope of 0.175 per cent. If the EBOD were 600 mg/l instead of
500, the effective slope for the Curve A condition at 2.0 cfs would be:

$$0.175 \text{ percent} \times \sqrt{600/500} = 0.19 \text{ percent}$$

For a system in which effective slopes are substantially higher than required by
Curve A, sulfide concentrations will be negligible for practical purposes. For
conditions substantially below Curve B, higher sulfide concentrations must be anticipated, generally requiring odor and corrosion control measures, or continuous treatment of the wastewater to prevent sulfide build-up."

The Manual makes further definite statements concerning the life expectancy of concrete sewers:

"If the slope-flow relationships of sewers upstream from a given point correspond to
Curve A (adjusted for EBOD), and there are no force mains operating without proper
sulfide control, then sulfide concentrations will be so low that the rate of corrosion of concrete pipe will be inconsequential. Small collecting sewers so designed,
made of concrete pipe with granitic or other inert aggregate will have a life expectancy of 100 years or more . . . Considering the thickness of the pipe wall in the
very large pipes, a life of several centuries would be expected. Where over-all
slopes are represented by Curve B, sulfide conditions under some circumstances may
be such that bare concrete pipe made with granitic aggregate will be significantly
corroded. Sulfide conditions become worse at slope-flow combinations deeper in the
domain below Curve B. It may be satisfactory to use concrete pipe under these conditions if it is made with calcareous aggregate. . . ."

QUANTITATIVE SULFIDE PREDICTION FORMULAS

Ref. 2 includes a quantitative predictive formula for full pipes, such as pressure (force) mains, inverted siphons or surcharged gravity sewers. The formula is:

$$\frac{d(s)}{dt} = 1 \times 10^{-3} \, (EBOD) \, r^{-1} \, (1 + 0.37D)$$

where:

$$\frac{d(s)}{dt} = \text{sulfide build-up rate mg/l-hr.}$$

$$EBOD = \text{Effective BOD mg/l}$$

$$r = \text{Hydraulic radius, meters}$$

$$D = \text{Pipe diameter, meters}$$

A quantitative sulfide prediction formula for gravity sewers was published by Pomeroy and Parkhurst (Ref. 10) in 1977. The formula is:

$$\frac{d(s)}{dt} = M' \, (EBOD) \, r^{-1} - N(su)^{3/8}$$

where:

$$\frac{d(s)}{dt} = \text{sulfide build-up rate mg/l-hr.}$$

$$EBOD = \text{Effective BOD mg/l}$$

$$r = \text{hydraulic radius, meters}$$

$$s = \text{slope of energy line}$$

$$u = \text{velocity of the stream, m/sec.}$$

$$M' = \text{Coefficient for sulfide build-up term}$$

$$N = \text{Coefficient for sulfide loss term}$$

Values for M' and N should be selected that will show what build-up may occur when all of the factors are in a range favorable for build-up to occur. The paper suggests two sets of values, "moderately conservative" with $M' = 0.32 \times 10^{-3}$ and N = 0.96 and "more conservative" with $M' = 0.32 \times 10^{-3}$ and N = 0.64. The "moderately conservative" coefficients have been suggested for use by the author (Ref. 11) when evaluating a series of reaches of a long interceptor. Adequate data to predict sulfide loss or dissipation at structures is not available. It is extremely unlikely that each successive reach of an interceptor sewer would generate sulfides at the maximum predicted level. To make this assumption and pass that sulfide level on to each succeeding reach would be unreasonable, particularly since the losses at structures are ignored.

Recent research has pointed up the fact that it is valid to use the "moderately conservative" coefficients when evaluating longer lines. This is a rather remarkable finding when one considers that the coefficients were determined based on studies of relatively short reaches, representing from 30 minute to as short as 15 minute flow times, with sodium sulfide added at upstream manholes so that changes could be determined at various sulfide concentrations.

THE DELANO, CALIFORNIA STUDY

A study by Warren and Tchbanoglous (Ref. 12) of two interceptors in Delano, California found that:

"An excellent correlation between the calculated and the measured sulfide levels was obtained on the first trial" and that "The calculated and actual average total sulfide values for the six runs were identical."

Values of M' = 0.35×10^{-3} and N of 0.6 were used for the prediction.

Attempts to correlate the observed sulfide build-up with the Thistlethwayte prediction formula showed "essentially no correlation" between predicted and observed values. The tests were conducted on the Stradley Avenue trunk, an 18" diameter vitrified clay pipe line at a slope of .0008. The length of sewer covered in the tests was 2,800 feet with transit times of approximately 30 minutes. The study states further that:

"The sulfide levels found in the Delano system could have been <u>qualitatively</u> predicted by use of Pomeroy's flow-slope relationship." (Fig. 4)

"By definition, sewers having values on or above Curve A have a very slight sulfide producing potential, while sewers with values on or below Curve B have definite sulfide producing potential. As seen, the Cecil Avenue trunk value is above Curve A (no sulfides) and the Stradley Avenue trunk is below Curve B (sulfide generation)."

SACRAMENTO, CALIFORNIA CENTRAL TRUNK STUDY

A second study (Ref. 13) recently completed by Gilbert Associates, a division of Brown and Caldwell, examined sulfide build-up along an entire 21 km (13 mile) long concrete pipe interceptor sewer known as the Central Trunk Sewer. The interceptor varied in size from .69 m to 1.52 m (27 inches to 60 inches). Severe sulfide problems were encountered in the Central Trunk and the Sacramento County Sanitation District collected extensive data on the interceptor over a period of eleven years.

The study evaluated both Thistlethwayte and Pomeroy-Parkhurst sulfide prediction formulas. When the entire 21 km (13 mile) interceptor was considered, the Pomeroy-Parkhurst formulas with an "N" of .96 (moderately conservative) was found to predict sulfide levels approximately twice those observed (Fig. 5). When the effect of dilution at junctions was taken into account (by a mass balance at each major contributing flow point) the resulting sulfide build-up predicted (Fig. 5) closely approximates the observed build-up.

It appears reasonable then, when evaluating long interceptor systems (one hour flow times or longer), to use the Pomeroy-Parkhurst formulas with the less conservative coefficients. The resulting equations become:

$$\frac{d(s)}{dt} = 0.32 \times 10^{-3} \ (EBOD) \ r^{-1} - 0.96 \ (su)^{3/8} D^{-1} (s) \ \text{mg/1-hr.}$$

$$S_\infty = \frac{0.33 \times 10^{-3} \ (EBOD)}{(su)^{3/8}} \times \frac{P}{b} \ \text{mg/1}$$

$$S_2 = S_\infty - \frac{S_\infty - S_1}{\log^{-1} \frac{(su)^{3/8}}{2.303 \ d_m}} \ \text{mg/1}$$

Where:

S_∞ = limiting sulfide level, at which sulfide build-up and losses are equal. (If the input sulfides are <u>less</u> than S_∞, they will <u>increase</u> toward S_∞. If input sulfides are <u>greater</u> than S_∞, they will <u>decrease</u> toward S_∞.)

D = pipe diameter, meters

d_m = mean depth, meters. (Area of flow divided by width of flow.)

All other variables are as previously defined.

Ref. 11 lists the above equations for use with Imperial units. The article includes a computer program for small desk top programmable calculators. "Fortran" and "Basic" language programs are available.

SULFIDE PREDICTIVE STUDIES ARE AN IMPORTANT PART OF THE DESIGN PROCESS

Sulfide prediction studies are recognized as an important phase of the initial design contract and as such are eligible for Federal funding by the USEPA.

The sulfide study (Ref. 15) which was performed for the 200 million pound (400 million dollar) Sacramento Regional System was a 160,000 pound (320,000 dollar) design study. From the results of the study, sulfide control methods can be designed into the system and/or appropriate pipe protection steps taken to assure the desired service life. For portions of the Sacramento system, oxygen injection directly into the wastewater was used for sulfide control. Other commonly used methods include chlorination, hydrogen peroxide injection, aeration, periodic "shock" treatment with caustic soda (some of which is available as an industrial waste), iron salts and nitrates. Ref. 2 includes an extensive discussion of sulfide control methods, including several case histories.

Oxygen injection into wastewater streams is planned for a project in the Middle East in the Emirate of Sharjah. An advantage to oxygen enrichment in the sewer is the increased biological activity that takes place in the sewer system prior to reaching the treatment plant. This feature reduced the required capacity in the activated sludge aeration tanks. This saving alone was nearly enough to pay for the oxygen injection equipment.

Sulfide studies are currently being performed by the consultants on two of the major sanitary sewer projects in the world -- the Manila, Phillipines system and the Sao Paulo, Brazil system. The Sao Paulo project includes the collecting system and treatment system for a 5,680 million litres/day (1,500 mgd) facility.

In order to control sulfides or design pipe for their effects, an accurate method of sulfide prediction is needed. In the past 35 years, we have seen continual upgrading of qualitative sulfide criteria. The Pomeroy-Parkhurst formulas, with mass balance techniques applied at junctions, appears to provide a reliable method for predicting sulfides in sanitary sewer systems.

DESIGNING CORROSION RESISTANCE INTO SEWERS

Where complete sulfide control cannot be achieved, methods have been developed to design for a particular service life under corrosive conditions. In the past, additional concrete thickness, calcareous aggregates or combinations of the two have been used to extend the life of concrete pipe sewers in a sulfide environment. Additional cover and alkalinity requirements were based on judgment and experience. Too often, the results were an excessive amount of protection at greatly increased costs.

Until the publication of the EPA Manual (Ref. 2), there was no rational method available to calculate the required alkalinity and/or the amount of additional concrete thickness needed. The EPA Manual relates those factors which affect H_2S transferred to the pipe wall to corrosion rate and alkalinity. The equation expresses average rate of penetration of the pipe material in inches per year, based on flow conditions, sulfide levels, pH and the alkalinity of the material as follows:

$$c = \frac{11.5 \, k\phi_{sw}}{A}$$

Where: c = corrosion rate, mm/year

     k = < 1.0, all losses in reaction

     ϕ_{sw} = sulfide flux to the pipe wall gm/m^2-hr.

     A = concrete alkalinity

The author, with Dr. Pomeroy, published a paper (Ref. 16) which presents design methods for other pipe materials including asbestos-cement pipe, mortar-lined ferrous pipe and ductile iron pipe under sulfide corrosion conditions.

DISTRIBUTION OF CORROSION EFFECTS

Ref. 2 states that the point of maximum corrosion can be as much as 1.5 times the average corrosion "c" calculated by the above EPA formula. The corrosion is not always uniform, with the area of accelerated corrosion occurring at the crown. The average rate "c" exists within the pipe where relatively stable conditions exist. Within one pipe diameter or so of the manhole, the downstream pipe will sometimes exhibit markedly increased corrosion if the structure is poorly designed and much turbulence occurs. The increased corrosion rate immediately adjacent to the manhole in severe cases may be several times the calculated average rate "c".

CENTRAL TRUNK STUDY VERIFIED EPA CORROSION RATE FORMULA

One of the objectives of the Sacramento Central Trunk Study (Ref. 13) was to further verify the EPA corrosion rate formula. The observed corrosion in the 21 km (13 mile) interceptor was compared with the calculated average corrosion rate "c". Table 1 lists the predicted and observed values of corrosion at the manholes. The observed corrosion varied from 25% to 150% of the calculated average. The worst corrosion (at twice the calculated average corrosion) occurred immediately downstream of a structure in which extreme turbulence was generated.

Interestingly, if the Pomeroy-Parkhurst sulfide prediction formula (less conservative) coefficients was used to predict sulfides and a factor of safety of two on design life was applied, the protection designed into the pipe would still have been twice that required even at the most severely corroded point.

LIFE FACTOR DESIGN METHOD

In 1974, the author suggested a design method (Ref. 17) which combined the principle of extra concrete thickness and increased concrete alkalinity. The method is referred to as the "Az factor" or "Life factor" method of concrete sewer pipe design. A "life" equation can be written as:

$$L = \frac{z}{c} \qquad \text{where}$$

L = required design life in years

z = available material thickness for corrosion protection, mm.

c = corrosion rate, mm/yr

Substituting the EPA Manual equation for corrosion rate "c" into the "Life" equation yields:

$$\text{Life } L = \frac{Az}{11.5 \, k\phi_{sw}} \qquad \text{(yrs.)} \qquad \text{Rearranging terms gives:}$$

$$11.5 k\phi_{sw} L = Az \quad \text{(mm)}$$

The term "Az" is called the "Life Factor" or "Az Factor" and is the product of concrete alkalinity "A" times the thickness of allowable concrete loss "z".

Concrete alkalinity is defined in terms of the amount of acid which a given weight of the concrete can neutralize as compared to the acid neutralized by the same weight of pure $CaCO_3$. The author has recommended that a factor of safety of 2 be applied to design life and that for reinforced concrete pipe, the "z" should represent the thickness of cover over the reinforcing steel at the crown of the pipe (Fig. 6). The revised life factor equation (with a factor of safety of 2.0) becomes:

$$23k \, \phi_{sw} \, L = Az \text{ (mm)}$$

The calculated "Az" is specified by the design engineer for each reach of the sewer
line. Ref. 16 includes design charts from which the design "Az" values can be directly determined.

This design procedure was first used in the Herndon Avenue interceptor project in
Fresno, California in 1974. The project, designed by Braun, Pasillas and Wagner,
Consulting Engineers, consisted of 27.4 km (17 miles) of 1 m through 2.13 m (39"
through 84") sewer pipe. The specified "Az" values ranged from 17.8 mm to 35.56 mm
(0.70" to 1.40"). By way of comparison, conventional granitic aggregate concrete pipe
with cover over the steel of 19mm (3/4") will have an "Az" factor of about .16 x 3/4"
equals .12 inches (3.05 mm). The specified life factor pipe would have a life expectancy of from six to twelve times that of conventional concrete pipe.

The advantage of the "life factor" method is the flexibility that it permits the pipe
manufacturer in producing a pipe that meets the desired corrosion resistance level.
Each manufacturer can use that combination of concrete mix, aggregates, wall thickness, steel area and concrete cover that is most economical for his particular production equipment, forms, aggregate sources and costs, etc.

Quality control for Life factor pipe is quite simple. The procedure consists of obtaining two one inch drill hole samples of the cover concrete from each test pipe
(Fig. 7). A standard $CaCO_3$ equivalent alkalinity test is run on each sample and
the results reported as the average of the two tests. The reported "Az" for the
test pipe is the thickness of concrete cover in mm over the reinforcing steel
multiplied times the alkalinity.

Significant cost savings can be realized through "life factor" pipe design. To date
over 161 km (100 miles) of "life factor" pipe have been designed and installed in the
United States. Savings over alternate products, primarily PVC lined concrete pipe,
vitrified clay pipe, etc., have been as much as 25% of the project cost. "Life
factor" design is being studied by Metcalf and Eddy as an alternate material for the
Sao Paulo, Brazil Regional Interceptor/Treatment Project. Montgomery Engineers will
use the principles of "Life Factor" design in the Manila, Phillipines Regional Project.

The City of Los Angeles, in a recent study of cost effective alternatives, estimated a
750,000 pound (1.5 million dollar) savings could be realized by using "life factor"
pipe as an alternate to vitrified clay pipe and PVC lined concrete pipe. The savings
represented 36% of the estimated project pipe costs.

One of the most interesting aspects of the method is that each test of pipe material
is an actual PERFORMANCE test, in which one normal hydrochloric acid is used to determine the acid neutralizing capability of the concrete protecting the reinforcing
steel. Each test pipe is a "case history" of its own.

The methods available to the sanitary design engineer today insure that cost effective
designs which will provide the desired service life can be calculated, specified and
constructed. Sulfide build-up can be accurately predicted and control methods designed to eliminate or reduce the hazard. There were two recognized alternatives in
the past -- protected concrete pipe and unprotected concrete pipe. Too often the
degree of protection was far in excess of what was needed for the job.

With the current "State of the Art" the high percentage of successful use of concrete
pipe for sanitary sewers should become even higher.

****

The author wishes to thank Walt Meyer, J. B. Gilbert & Associates, a Division of Brown
& Caldwell, for providing information for Ref. 13 which, at the time of writing, had
not yet been published.

REFERENCES

(1) Pomeroy, R. D. and Bowlus, F. D., *Progress Report on Sulfide Control Research*, Sewage Works Journal, 18, No. 4, pp 597-640 (1946).

(2) *Process Design Manual for Sulfide Control in Sanitary Sewerage Systems*, U. S. Environmental Protection Agency (October, 1974).

(3) *Report and Recommendation of Regional Counsel to the Regional Administrator*, Contract No. C-06-1007, USEPA Region IX, San Francisco, CA (1977).

(4) Davy, W. J., *Influence of Velocity on Sulfide Generation in Sewers*, Sewage and Industrial Wastes, 22, No. 9, pp 1132-1137 (1950).

(5) Pomeroy, R. D., *Sanitary Sewer Design for Hydrogen Sulfide Control*, Public Works (October, 1970).

(6) Thistlethwayte, D. K. B. (Editor), *Control of Sulphides in Sewerage Systems*, Butterworths Pty. Ltd., Melbourne, Australia (1972) and Ann Arbor Science Publishers, Ann Arbor, MI (1972).

(7) Parker, C. D., et al., *Concrete Sewer Corrosion by Hydrogen Sulphide*, Melbourne and Metropolitan Board of Works Technical Paper No. A.8, Part 2, Generation of Sulphides in Sewers.

(8) Pomeroy, R. D. and Parkhurst, J. D., *Self-purification in Sewers*, Proceedings of the 6th International Conference on Water Pollution Control Research, Jerusalem, Pergammon Press (June, 1972).

(9) Pomeroy, R. D., *Generation and Control of Sulfide in Filled Pipe*, Sewage and Industrial Wastes, 31, No. 9, pp 1082-1095 (1959).

(10) Pomeroy, R. D. and Parkhurst, J. D., *The Forecasting of Sulfide Build-up Rates in Sewers*, Prog. Wat. Tech., Vol. 9, pp 621-628, Pergammon Press, Printed in Great Britain (1977).

(11) Kienow, K. K. and Kienow, K. E., *Sulfide Prediction/Concrete Pipe Interceptor Sewer Design Program for the HP-67/97*, Water & Sewage Works (September, 1978).

(12) Warren II, G. D. and Tchobanoglous, G., *A Study of the Use of Concrete Pipe for Trunk Sewers in the City of Delano, California*, Department of Civil Engineering, University of California at Davis, California (September, 1976).

(13) Meyer, W. J. and Hall, G. H., *Prediction of Sulfide Generation and Corrosion in Concrete Gravity Sewers, A Case Study*, J. B. Gilbert & Associates, A Division of Brown & Caldwell (April, 1979).

(14) Pomeroy, R. D., *The Problem of Hydrogen Sulfide in Sewers*, Clay Pipe Development Association Ltd., London (1978).

(15) *Control of Odors and Corrosion in the Sacramento Regional Wastewater Conveyance System*, Sacramento Area Consultants, Sacramento, CA (September, 1976).

(16) Kienow, K. K. and Pomeroy, R. D., *Corrosion Resistant Design of Sanitary Sewer Pipe*, ASCE Convention and Exposition, Chicago (October, 1978) Water & Sewage Works, Reference Issue (April, 1979).

(17) Kienow, K. K., *Protecting Reinforced Concrete Sewers*, Water & Sewage Works, Vol. 122, No. 10 (October, 1975).

TABLE 1

COMPARISON OF PREDICTED AND OBSERVED PIPE CORROSION
(from Ref. 13)

Inches	Pipe Size meters	Predicted Corrosion[1] inches	mm	Measured Corrosion[2] inches	mm
27	.69	.1	2.5	0	0
30	.76	.2	5.0	0	0
36	.91	.2	5.0	.3	7.5
39	1.0	.3	7.5	.3	7.5
42	1.07	.4	10.0	.6	15.0
48	1.22	.6	15.0	.8	20.0
54	1.37	.7	18.0	.7	18.0
60	1.52	.8	20.0	.8	20.0

[1] Average Corrosion, $k = 1.0$, using EPA formula $c = 11.5k\phi_{sw}\frac{1}{A}$ mm/yr.

[2] Measured crown corrosion at downstream manhole.

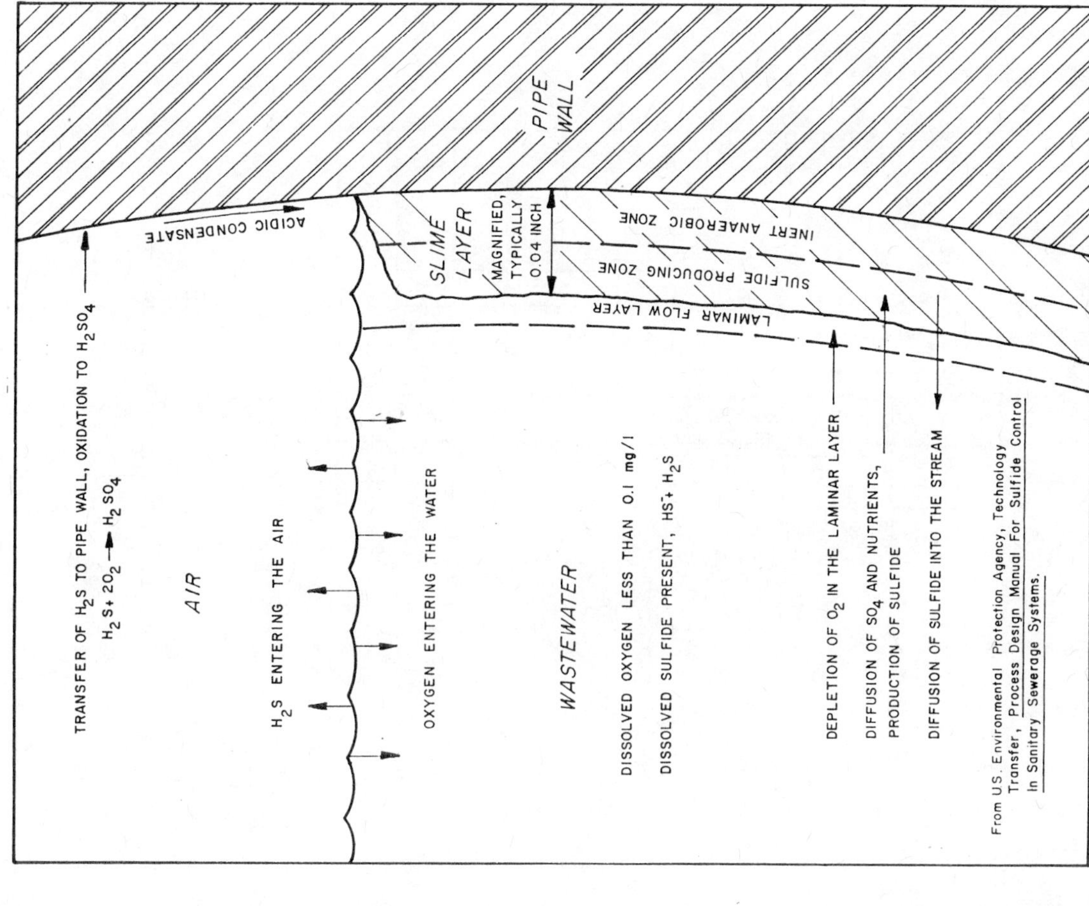

Fig. 2. Processes occurring in sewers under sulfide buildup conditions

Fig. 1. Processes occurring in sewers with sufficient oxygen to prevent sulfide from entering the stream

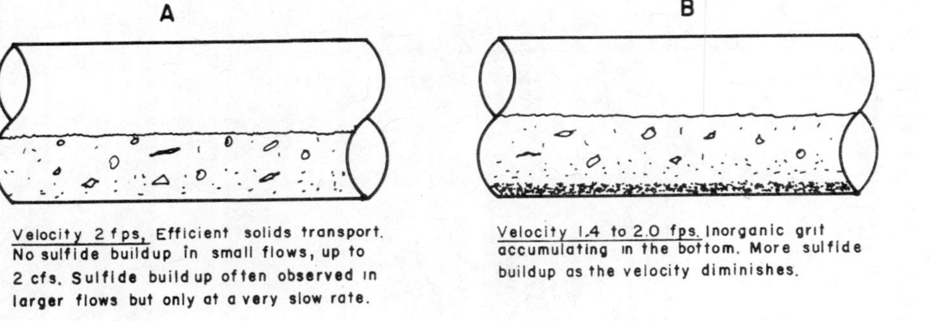

Fig. 3. Solids accumulations at various flow velocities

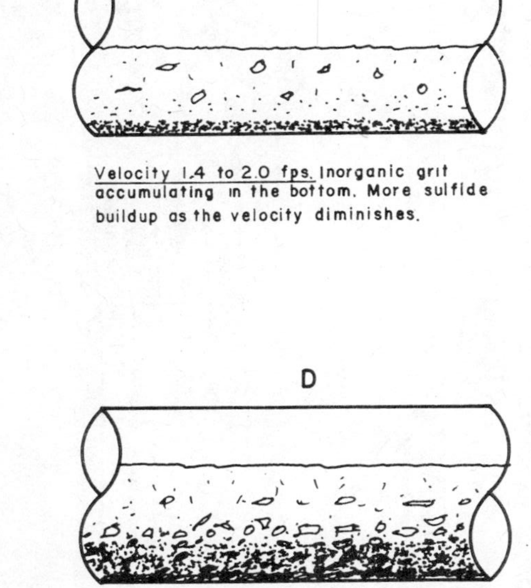

Fig. 4. Flow-slope relationships as guides to sulfide forecasting

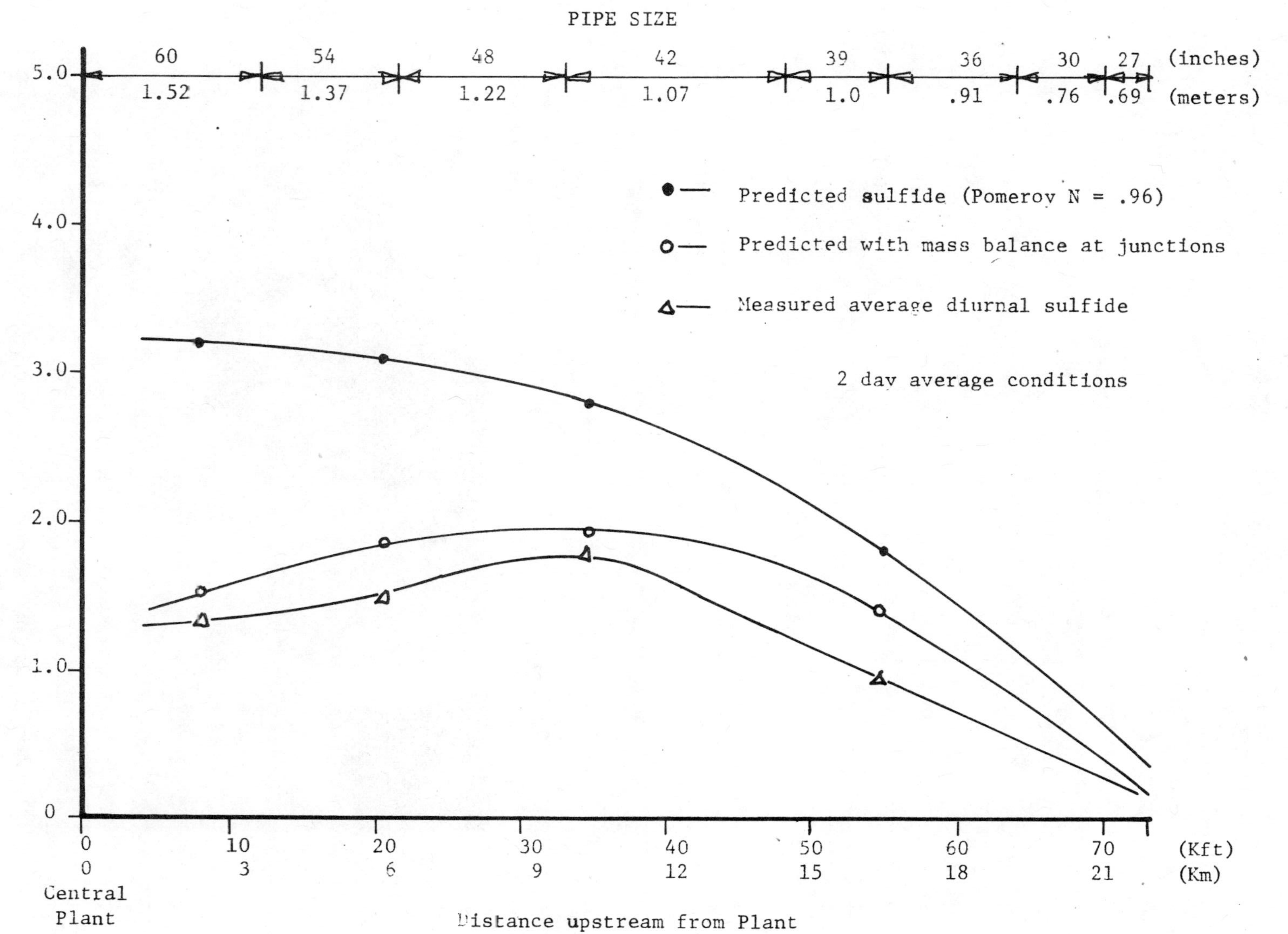

Fig. 5. Predicted and observed sulfides in the central trunk

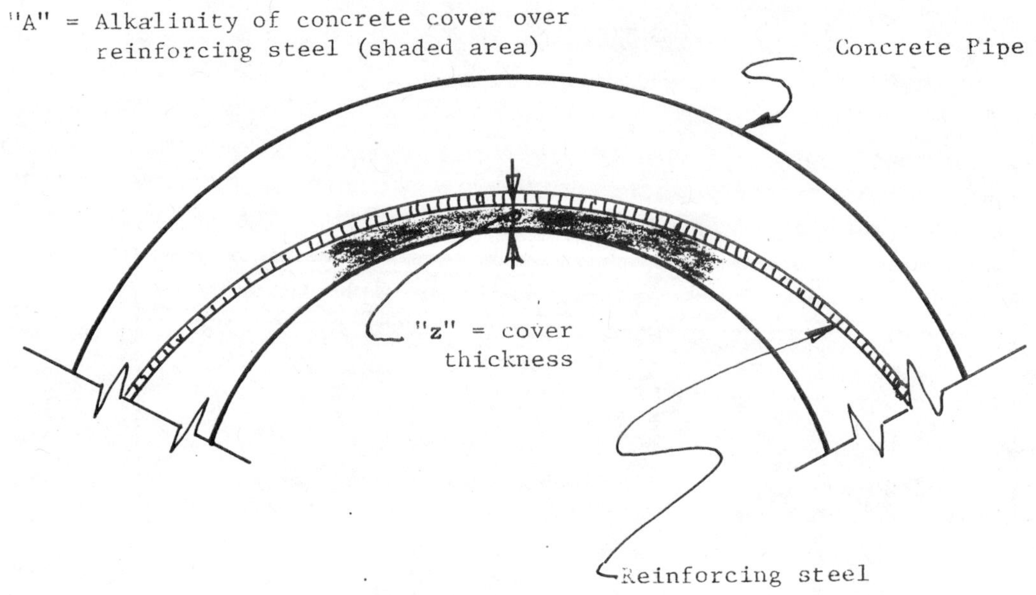

Fig. 6. "Az" design concept

Fig. 7. Drilling sample for alkalinity testing

internal and external

protection of pipes

September 5th - 7th, 1979

USE OF POLYAMIDE 11 AS PIPE LINER FOR SLURRY PIPELINE

R.A. Ganga

ATO Chimie, France

V.T. Nguyen

Omnium Technique Des Transports Par Pipelines, France

Summary

Transporting pulverulent materials by pipes in a slurry form frequently generates the joint phenomena of corrosion and abrasion.

According to the characteristics of the transported materials to their action on the carrying fluid, the effects are of various importance.

The lining of pipes with an organic thermoplastic material can solve such problems.

Here is a study carried out by the O.T.P. team with a specific product to be transported. The test consists of pumping the slurry through samples lined with polyamide 11. The following are discussed: Presentation of polyamide 11; Pipe lining method; Solving the pipe junction problem; Laboratory tests; Description of the test facility; Preparation of samples; Definition of the product; Results.

Held at Imperial College, London, England.
Organised and Sponsored by BHRA Fluid Engineering

1. Presentation of Polyamide 11

Among all existing organic plastics   polyamide 11, also called
"nylon 11" over the Atlantic Ocean, has a particular importance
since its basic product is not derived from oil but from a vegetable,
the castor oil.

The polycondensation of the amino-undecanoïc acid, which is the basic
molecule, allows, thanks to chain limitators, a state of polymerisation
varying according to the intended application to be obtained.

The formula $H_2N - (CH_2)_5 - COOH$
shows indeed that blocking in such or such direction will give basic
or acid reactions. Depending on the chemical composition of the
transported liquid, the one or the other function will be used for
the product grade developed.

The resulting polymer is thus at equilibrium

- with itself (but the polycondensation is blocked by the chain limitator)

- with the water it contains.

A polyamide will absorb all the more water as the number of amide groups
are larger for the same weight ; in other words as the number of CH_2
groups will be smaller.

Consequently   polyamides 11 are those which absorb the smallest
possible quantity of water, which results in :

- an excellent dimensional stability in time

- stable electrical (conductive) or mechanical (plasticizing) properties,
  hence

- good ageing properties.

As compared with other polyamides, polyamide 11 has -because of its
chemical composition- the following characteristics :

- a lower density (1.05)

- a lower melting point (186° C)

- is less rigid

- is less brittle under impact.

It is also possible by means of a plasticizer to reduce the intermolecular
links in order to get a comparatively more flexible product which might be
interesting for transporting solids when the speed of the particles demands
bouncing so as to reduce wear.

Taking account of these different elements, various grades of polyamide
have been developed for answering specific requirements such as

> - transported products

> - mode of application of the product.

The carrying fluid most frequently used is generally water with pH varying
according to the product in suspension.

In order to put in evidence the polyamide 11 resistance in an aqueous environment, the following tests have been carried out :

- oxygen resistance

- capillary wall action

- auto-corrosion

- vapour test

- boiling water

- thermal shock

(water pH : 4.5)

The above tests have led to the development of a special grade for hot water which is used at present for the internal coating of boilers (10 years warranty). This has enabled the suppression of enamelling as well as the cathodic sacrificial magnesium anode. This "hot water" grade is also used for coating internally sea-water-conduits feeding refrigerating stations in which turbulence of loaded water is critical. So, the piping is able to operate for 40,000 hours against 1,000 formerly. (cf. report of the Compagnie Générale Transatlantique).

Special grades are also developed according to the mode of application of the material (which in the case of coatings is always in the form of powders).

The well known processes of

- electrostatic gun spraying and

- dipping, which allow only thin coatings from 100 to 600 microns

should be avoided in the case of abrasive products.

Thick coatings are recommended in order to comply with severe long life requirements.

Grades have been specially developed for rotational moulding. The additives they contain do not modify in any way the abrasion resistance of the product, which is one of the main characteristics of polyamide 11. As a matter of fact, tests of the basic polyamide carried out with a TABER abrasimeter indicate after 1,000 cycles under 1,000 grams a loss of weight of 5 mg only.

Even if this indication cannot be extrapolated to products in suspension, it is an excellent means of comparison with other plastics.

2. Application of the Material

For correct application of the polyamide it is necessary to prepare the surface of the substrate and to use a primer or undercoat that will allow an excellent adhesion, essential to non corrosion.

Since it is a thermoplastic material, the application consists of a single fusion at the required temperature, which gives an even coating with the same resistance for all grains : this is the main criterium of resistance to abrasion.

a) <u>Surface preparation</u>

    For steel piping **securing** should be performed with angular shot. The
surface state obtained complies with the Swedish standard : 2.5 (or
with the American Std. 10 NEAR WHITE).

b) <u>Primer coating</u>

    A RILPRIM 104 primer is applied as a liquid on the cold tube by means
of a spray-gun so as to obtain a dry film of about 10 microns. The
primer dries in 3-4 minutes and resists heat. It affects in no way the
ulterior processing of the polyamide for it does not decay when passed
over the flame.

c) <u>Application of the polyamide</u>

    The powder necessary to the final coating is filled inside the tube
in a quantity corresponding to the required thickness (1.05 kg/mm/m2).

    Either ends of the tube are fitted with caps providing a spare space
of about 8 cm enabling **the tube to be post-weld without causing any**
damage to the coating. Then the tube rotates at a speed of 4 m/min.
and is heated by means of a gas nozzle the size of which depends on
thickness and diameter of the tube to be coated.

    Fusion takes place slowly and continuously, thus giving a homogeneous
coating.

    The heat diffusion through metal is sufficient to ensure fusion
without overheating at temperatures between 220 and 240° C.

    A visual examination is possible by using white powder the immediate
yellowing of which indicates any deterioration of the product. This
simple quality control serves as a criterium for a good application of
the thermoplastic.

    See fig. 1.

3. <u>Problem of the unions (or joints)</u>

    Contrarily to thermosetting materials the thermoplastic polyamide allows
an easy touch-up of the coating at the joints.

    Thanks to the 8 cm spare space, welding is possible without damaging the
thick coating.

    An open ring narrower but thicker than the spare space is then inserted
in the tube after welding so as to fit on the spare section. The ring
is applied to the tube walls by means of an expansion machine.

    The tube is heated externally in an annular gas furnace enabling both
the initial coating and the inserted ring to fuse. When melting, the
material inside the tube gives an even coating without any discontinuity
between the initial coating and the inserted ring.

    This type of union is easier to carry out for large diameter tubes where a
visual control by inserting a manual applicator is possible.

    For small diameter tubes another technique has been developed which is
subject to a TRIWELD patent held by SPIE, Batignolles. This technique
consists in welding -before carrying out the internal coating- a corrosion
(or abrasion) resistant metal ring at the end of each tube ; this ring is

about 8 cm long. Then the coating is applied as a function of the thickness
required.

The two rings of special metal will be then simply welded on the field,
at the same working cadency as for usual steel welding.

See fig. 2

4. Abrasion tests

In order to appreciate the abrasion resistance of the polyamide 11, a larger
test programme was carried out.

The test consists of pumping an highly abrasive slurry through a test loop
on which samples are mounted. Samples are made with 2 inch   pipe sections
lined with polyamide 11.

The test runs took more than four hundreds hours. Various flow velocities
have been selected in order to find out the abrasion rate as a function of
flow velocities.

4.1 Test facilities

A 2 inch   test loop of 50 meters length was used for the tests.

Various samples - steel pipe section lined with polyamide 11 - are
mounted in series on the loop. In order to compare the abrasion
resistance of the Rilsan with that of steel, non lined steel samples
are also used.

The characteristics of the samples are :

- diameter   50 mm
- length   75 mm
- weight   up to 200 g

The various flow velocities selected are :

V_1 4 m/s
V_2 7 m/s
V_3 9 m/s
V_4 12 m/s

The slurry used is corrosive (pH $\simeq$ 4.5) and highly abrasive due to
its sand content (58 % in weight).

4.2 Abrasion rate measurements

The abrasion of the lining material during the tests involves a loss
of weight. The abrasion rate U (weight losses) is in the following
form

$$U = f(t,v) \qquad (1)$$

as far as the slurry characteristics (granulometry, concentration,
composition of solid.....) are concerned, these are unchanged.

For the same test time t, U will be a function of V

$$U = k\, V^{\alpha} \qquad (2)$$

U is measured by weighing. Relation (2) gives α.

5. Test results and conclusion

At the end of a 500 hundred hour test programme, the following conclusion
can be rised out :

- the polyamide 11 has an excellent resistance to abrasion as good as that
  of rubber at flow velocities up to about 7 m/s.

- for the flow velocities higher than 7 m/s, the abrasion rate increases
  and becomes about twice than that expected.

It seems that this is due to the relative hardness of the Rilsan which
is not suitable for high velocities.

Due to various reasons (head losses i.e. energy consumption, abrasion,
etc...) flow velocities do not exceed 2.5 meters per second in long distance
pipelines. This makes the polyamide 11 liners suitable.

The problem to be solved now is the junction method, avoiding if possible
the heating of the steel pipe to a high temperature.

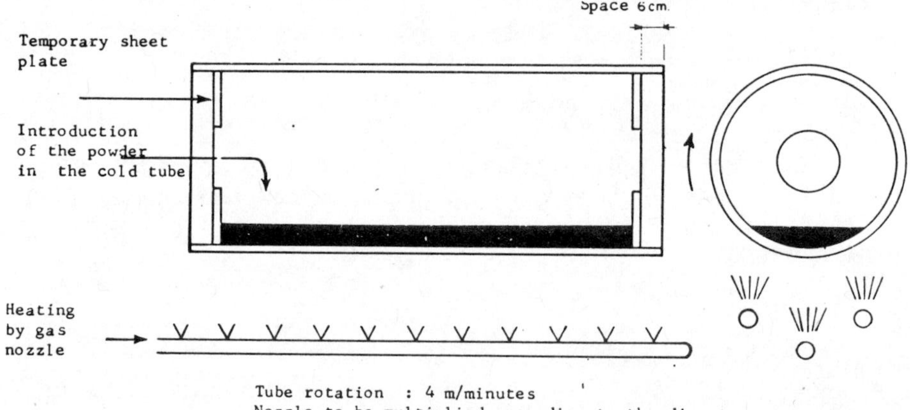

Fig. 1

Fig. 2

SAFETY EVALUATION OF PIPELINE

<u>D.J. Bryce BSc, PhD, and M.J. Turner MI Mech E.</u>

Health & Safety Executive, U.K.

Summary

This paper describes how the hazards and risks presented by a cross-country pipeline to populations or installations in the vicinity may be evaluated. Attention is given primarily to pipelines that carry flammable materials, but in principle the approach is considered equally applicable to any that carry toxic materials. Brief details of recent significant pipeline incidents which involved flammable materials are given.

It is emphasised that appraisal of the dangers depends on the hazardous nature of the material transported and on the risk or probability that it will be released to atmosphere. The hazards from flammable material are those of thermal radiation in the event of fire, or blast in the event of explosion. Flammable material will be ignited only if it comes into contact with an ignition source while mixed with air in a concentration between the flammable limits. The significance of the mode of emission from the pipeline and the processes of atmospheric dispersion in reducing the concentration of released material to below the lower flammability limit are described. Prediction of the risk of pipeline failure can be based on experience of existing pipeline systems, and the applicability of data from the US and European petroleum pipeline networks and the UK high-pressure gas networks are discussed.

Finally, the paper shows how information on rates of release and probable atmospheric dispersion may be combined with pipeline failure rate data to predict the frequency with which populations nearby are likely to be affected by a pipeline. The approach is illustrated by reference to a recent Health and Safety Executive pipeline safety evaluation.

Held at Imperial College, London, England.
Organised and Sponsored by BHRA Fluid Engineering

<u>NOMENCLATURE</u>

D   –   Distance between pipeline and population under consideration –m

P   –   Length of population parallel to pipelines –m

R suffix   –   Rates of release from pipeline in kg s^{-1}

D suffix   –   Dispersal distance to given concentration –m

P suffix   –   Length of pipeline that can affect population (pipeline interaction length) –m

F suffix   –   Frequency of occurrence of given rate of release in km^{-1} yr^{-1}

f suffix   –   Frequency of occurrence of leak rate over pipeline interaction length equal to F suffix x P suffix – yr^{-1}

f' suffix   –   f suffix modified by direction factor, (wf) –yr^{-1}

f" suffix   –   f' suffix modified by frequency of weather category (WF) to give interaction frequency –yr^{-1}

Where Suffix: 1, 2 and 3 refer to rates of release and A to G refer to category of weather

wf   –   Factor to allow for the probability of material dispersing in the direction of population

WF   –   Factor to allow for the probability of occurrence of weather category

1. <u>INTRODUCTION</u>

A safety evaluation by the Health and Safety Executive on two proposed
cross-country pipelines carrying flammable materials was given in a Report
(Reference 1) issued in July 1978.   This paper outlines the methods used
in that Report to assess and quantify the potential hazards and risks which
it is suggested can be used and extended to other pipeline situations
including those where toxic materials are carried.

The potential for a pipeline installation to cause harm or damage depends
upon the nature and quantity of the material that can be assumed to escape
from the pipeline and its subsequent behaviour.   The risk associated with
a pipeline depends upon the probability of such escapes and the likelihood of
the particular consequences.   The distinction between the evaluation of the
hazard and the risk is an important one to make and is reflected in the
layout of this paper, in that Section 2 discusses the hazard potential,
Section 3 the potential causes and statistics of pipeline failure, and Section 4 &
Section 5 bring the two aspects together to illustrate how (with examples
drawn from Reference 1) populations in the vicinity might be affected by a
given pipeline installation.

That a pipeline can be a potential threat has been demonstrated by a number
of significant pipeline incidents, where often considerable quantities of
flammable materials have been released, followed by fire or explosion.   To
predict possible consequences in a given situation is often complex and
depends upon many factors, including the mode of release, the initial mixing
with the air and subsequent dispersion.   Similarly the probability of an
event and its consequences depends upon the pipeline design, how it is
constructed, operated and surveyed, and on such factors as the probability
of ignition and explosion.

The results of safety evaluation used in the right context, knowing the
assumption and limitations, can help in highlighting areas where improvement
in safety can be achieved and as a guide to the judgment on the acceptability
of a particular installation.

2. <u>THE POTENTIAL HAZARD</u>

Hazard is defined here as the potential to cause harm or damage.   To assess
this potential threat for a pipeline, a number of factors need to be
examined, principally including:

     i)   the inherent properties and nature of the material
        contained in the pipeline,

    ii)   the rate and quantity of the material that can escape,

   iii)   the atmospheric dispersion characteristics of the
        material released,

    iv)   the possible effects of the released material.

These are discussed in turn as follows:

2.1 <u>The Nature of the Contained Material</u>

The main consideration of this paper has been directed towards pipelines
containing flammable materials as these constitute the majority of cases,
although there seems to be no reason why the approach should not be applied
to pipelines carrying toxic materials.

The principal hazard from flammable materials is clearly that on escaping
from the containment of a pipeline they will mix with air and when in
concentrations between the flammable limits will, if ignited, burn or
explode. The material is more likely to burn than to explode if ignition
occurs in the open air soon after the initial escape to atmosphere, in
which case a fire will persist at the point of emission until the flammable
material is exhausted. Here, the main hazard arises from thermal
radiation damage to people and property nearby, which may be significant
at hundreds of metres from the pipeline. However, if ignition is delayed,
there is the possibility of an explosion as the flammable vapour may build
up in a confined space such as a duct or drain, or form a large vapour
cloud containing large amounts of material. It is apparent that if a
cloud of flammable material is ignited before the cloud reaches a nearby
population, the population is liable to be exposed to thermal radiation
or explosion blast effects, depending on the distances involved. A
potentially more serious situation will occur if the cloud reaches the
population before ignition takes place because this situation will
involve more direct exposure to flame, radiation and blast.

With regard to toxic materials, the effects are not easy to summarise
briefly because of the great number of factors involved. It need only
be noted here that effects may be acute and quick-acting or chronic and
slow-acting; they may be those of direct and apparent poisoning or they
may be those of less apparent, perhaps carcinogenic responses; they may
depend significantly on the route of entry to the human body; they may
depend on a complex relationship between concentration and duration of
exposure.

2.2   The Rate and Quantity of the Material that can Escape

This aspect of the hazard evaluation is linked to the physical properties
of the pipeline system. It is necessary to postulate certain hole sizes
or range of hole size, and the statistics of previous pipelines as discussed
in Section 3 below may be of help in determining the most appropriate size
from which to calculate the possible leakage rates. This in itself can
be a complex fluid dynamics problem, particularly with two-phase flow,
as would be the case with natural gas liquids. Such factors as the
pipeline diameter and lengths, system pressure and flow rates need to be
taken into account. The total quantity of material that can escape is
generally more straightforward and will depend upon the detection time of
the system for particular leakage rates and the distance between stop valves
and the time it takes to close these. In this context it should be
remembered that for typical petroleum products significant quantities of
material (amounting to 100 – 250 tonnes per km for, say, a 300 mm diameter
pipeline) can be contained in a pipeline, and for particularly hazardous
material the value of remotely actuated stop valves that can readily be
closed is an important consideration in assessing the potential consequences
of a leakage.

2.3   Atmospheric Dispersion of Materials Released from a Pipeline

Flammable or toxic materials released to atmosphere may be expected to
become non-hazardous only by dispersion in the atmosphere. An initially
flammable material will only become non-flammable after it has been
diluted to a concentration below the lower flammable limit by the processes
of atmospheric dispersion, a potentially toxic material will only become
non-toxic after equivalent dilution to below the lower limit of toxicity.

The rate of dispersion will be dependent on the particular atmospheric

conditions prevalent at the time, but regardless of these the atmospheric
concentrations of released material at a fixed point down-wind of the point
of emission in general will vary directly as the rate of release to
atmosphere.   Thus, dangerous concentrations of flammable or toxic materials
will reach a given population only from rates of release above a certain
threshold level.   The rate of release will depend directly on the size of
the hole in the pipeline wall through which the material is escaping.

The process of atmospheric dispersion begins to operate immediately material
is released to atmosphere.   Understanding of atmospheric dispersion was
originally developed to describe plumes of neutrally-buoyant material, of
density comparable to that of the surrounding atmosphere, released from an
elevated source above a uniform terrain.   Such a plume tends to develop
down-wind in a roughly conical fashion, with dispersion occurring as a
result of turbulent motions within the plume.   The concentration at a
given point in the plume depends on the distance from the source and on the
degree of turbulent motion in the plume.   The degree of turbulent motion
depends on the weather.   Weather has been classified into seven categories,
Pasquill A to G (so-called after the founder of the classification system),
which respectively describe the range of conditions between a highly turbulent
summer's day and a still winter's night.   Concentration in a dispersing plume
will fall off more quickly with increasing distance from the source of
emission in turbulent conditions, more slowly with increasing distance in
still conditions.   Methods of predicting the dispersion of neutrally-buoyant
plumes become less reliable when the dispersion occurs over topography which
has significant features such as hills, open water, etc, because of the
interaction such features have with the atmosphere.

An important assumption in many calculations of atmospheric dispersion is
that the rate of emission will be less than the level at which turbulent jet
flow of the emitted material begins to entrain air from the surrounding
atmosphere.   This may not be a valid assumption in certain hole size ranges
for damage to pipelines.   Consideration may need to be given in a pipeline
hazard assessment to such effects, and perhaps to the results of turbulent
interaction with the walls of any trench etc, formed in the ground by the
emitted material, since either would decrease the initial concentration of
the released material and would in turn lead to a decreased distance to the
lower flammable or toxic level.

Wind plays a significant part in atmospheric dispersion, and to a first
approximation with neutrally-buoyant emissions it may be assumed that the
concentration at a fixed point down-wind of a source of emission will vary
inversely as the wind speed.   It is apparent that wind direction is important
in determining the concentration of emitted material received at a given point.

Another important aspect in determining the course of atmospheric dispersion
is the density of the emitted material relative to the surrounding atmosphere.
The established methods for estimating atmospheric dispersion deal with gases
with a specific gravity close to that of the surrounding air.   The methods
may not be used without special allowance for gases such as methane, which is
less dense than air and will tend to rise in the atmosphere, or for gases
such as propane or butane, which are more dense than air and will tend to
slump to the ground.   Dense gases released to atmosphere will slump to the
ground to form a low-lying pancake-shaped cloud which on a level surface may
on occasion flow against the wind, and which will follow surface gradients
down available valleys etc to collect on low lying ground.

Positive density effects may be affected by the nature of the material emitted
or by the circumstances of the emission.   Pipelines are found which transport

liquids held under pressure at temperatures above their boiling points.
Damage to such a pipeline would release liquid, which would immediately undergo
adiabatic flushing as its temperature fell to its boiling point under
atmospheric pressure. A cloud of vapour and suspended droplets would result.
The droplets would evaporate and in this evaporation would take up heat from
the entrained air. This would cool the air, and as a result a vapour—air
mixture might be formed with a net density greater than that of the surrounding
air, despite the fact that the material released from the pipeline had a
relative density less than that of air under normal conditions of temperature
and pressure. Such a mixture would disperse as a dense gas.

Alternatively, it is to be noted that conditions might exist with a dense
gas such as propane where the jet characteristics of an emission might be
such that sufficient air was entrained to give a resultant cloud with density
little different from the surrounding atmosphere which would thus disperse in
a neutrally—buoyant fashion.

In summary of this section, therefore, the concentration of airborne material
received at a population near a pipeline, in the event of leakage from the
pipeline, will:

    i) be larger with larger rates of release,

   ii) be larger with more stable weather conditions,

  iii) be smaller with increasing jet entrainment of air at the
        point of emission,

   iv) be dependent on the wind direction, unless the emitted
        material forms a cloud which is more dense than air, when
        surface gradients may be important,

    v) be smaller with larger wind speeds in situations where
        effects caused by the density of the emitted cloud are
        not important.

2.4   The Effects of the Released Material

The appraisal of the potential hazards from a pipeline depends upon consideration
of the material properties, the rate and quantity of the release and the
dispersion characteristics as discussed above, and the particular location.
For a flammable material, realisation of the hazards depends primarily on
the material being ignited. Ignition sources are most likely to be found near
to and among dwellings. To a first approximation, the frequency with which
the hazard will be realised depends upon the frequency with which vapours above
the lower flammable limit will reach the dwellings. This frequency depends
in turn upon the relationship between hole sizes that occur in pipelines and
their frequency of occurrence. These aspects are discussed more fully in
Section 3 below.

3.   POTENTIAL CAUSES OF PIPELINE FAILURE

Cross—country pipelines have failed over the years, sometimes in a spectacular
position, sometimes with fatal results. Outline details of a few incidents
are given in Table 1.

The general causes of failure fall into broad categories such as corrosion,
mechanical failure, operational error, third party activity and special hazards.
Each class needs to be recognised and evaluated in any pipeline hazard assessment.

Corrosion may occur within the pipe due to the chemical composition of the
fluids conveyed, or on the outer surface of the pipe due to the action of the
local environment.

The types of mechanical failure which may occur are perhaps primarily related
to the particular features of each pipeline. Incidents in the past have
occurred due to rupture of pressure transducer diaphragms, gasket failure on
pump station valves, failure of low pressure equipment wrongly subjected to main
line pressure, faulty welding procedures etc.

Operational error may occur in different forms. There may be situations where
it is possible to subject a pipeline to unacceptable stresses such as might arise
from fluid-hammer effects after the too rapid closing of a section isolation
valve or from thermal cycling. A different type of operational error has in
the past occurred when pipelines were changed from a low physical demand duty,
such as the transport of petroleum spirit, to a high physical demand duty, such
as the transport of liquefied petroleum gases, without sufficient thought being
given to the ability of the pipeline to meet the changed conditions satisfactorily. Operational error has also been seen in situations where pipelines,
on occasion with a history of failure, have been rushed back into service
after failure with as little delay as possible and with little regard for any
need to establish the cause of the failure.

Damage from third party activity is an ever-present problem, and includes the
type which occurs because of blows from mechanical implements used in civil
engineering works, agricultural draining operations etc. The damage is
seldom caused maliciously and is extremely difficult to guard against because
it happens largely through the actions of persons who are not aware of the
presence of the pipeline. Better protection can be obtained by, for example,
increased depth of soil cover, increased pipeline wall thickness, and more
adequate surveillance and monitoring. A feature of this type of damage in
the past has been the fact that a person striking a pipeline accidently with
a mechanical implement has often assumed no damage to have been caused, and
accordingly has not reported the incident to the pipeline operator. In fact,
such impact on the pipeline may easily damage the anti-corrosion cover, or
may set up unacceptable local stresses in the pipe steel, either of which
events may have significant delayed consequences.

The types of special hazard which may affect a pipeline clearly depend to a
very large extent on the geographical location of the pipeline and the precise
details of the ground through which it passes. Consideration may need to be
given to natural hazards such as geological activity, landslip, subsidence, or
peat and woodland fires, or to man-made hazards such as nearby pipelines or
crashing aircraft.

<u>Available Pipeline Failure Statistics</u>

Statistics on pipeline failure are clearly important in a pipeline hazard
assessment because they may be expected to give a guide to the probability of
material escaping from a pipeline to atmosphere and, in the general sense,
statistics from any sufficiently large and long-established pipeline system
would be useful. Difficulties, however, may arise in obtaining statistics
which are relevant to the precise situation on the particular pipeline for
which hazard assessment is required.

Such difficulties will arise in the need to refine statistics which relate
to a large pipeline network, such as that in the United States, until they
become suitable for application to the pipeline in question. It may be the
statistics include incidents which are not relevant. For example, if the
pipeline in question had no intermediate pumping station or was designed to
convey fluids with a sufficiently low water content, it would be necessary

to eliminate incidents which involved intermediate pump stations or internal
water corrosion from the network statistics before employing them in the
pipeline assessment.

Alternatively, difficulties may arise because statistics of the type in
question are historical in nature, and may well become out of date as a result
of technical advances or changes in procedures.   It might be that particularly
effective weld test methods when installing the pipeline were proposed or
that the pipeline was to be given an exceptional depth of soil cover, in which
case decisions might be taken to eliminate or reduce the number of incidents
concerned with weld failure or above-ground activities before using the
network statistics.   Equally, it is conceivable that an especially rigorous
or novel duty, such as transport of a cryogenic liquid,might require inclusion
of an increased failure rate in some aspect of the statistics.

From the British viewpoint, three extensive pipeline systems are apparent
from which statistics might be obtained.   These are the US petroleum system,
the European petroleum system and the UK national high-pressure gas transmission
system operated by the British Gas Corporation.

The pipeline systems in the United States are clearly a potential source of
a great deal of data as pipelines have been used there for more than 60 years.
However, there were no generally accepted standards or codes of practice until
the 1942 American Standard Code of Pressure Piping and, while revised codes
and standards of construction have appeared over the years, many pipelines
in use today were built to lower standards than are now acceptable.
Consequently, considerable effort in analysing the available data to identify
pipelines constructed and operated to standards comparable to those on any
pipeline being subjected to a hazard assessment may be necessary before the
US statistics can be adequately utilised, although a recent report (Reference 2)
by the National Transportation Safety Board (NTSB) provides a useful contribution to the statistics available from the United States.   Table 2 gives
some of the data given in that report.   The European petroleum pipeline
system, while not as old or extensive as the US system, may provide more
relevant statistics which can be employed in a pipeline hazard assessment.
Although constructed in places to differing national codes, the European
system is operated by multi-national companies which tend to have a common
approach and to employ common standards.   European experience of cross-
country pipelines is published annually by the oil companies group for the
Conservation of Clean Air and Water Europe (CONCAWE), but as the group's
title suggests, is concerned primarily with demonstrating the minimum effect
which the pipeline system has on the environment.   Table 3 gives a summary
of the CONCAWE data that was used in Reference 1.   The data is not reported
in the form which would be most useful in pipeline hazard assessment.
Prediction of the frequency with which a significant amount of material from
a pipeline will reach a nearby population depends on a knowledge of the
relationship between the hole sizes which are caused when damage is
experienced and the frequency with which each hole size range occurs, data
which is not given by CONCAWE.   (Thus it would be extremely useful if
CONCAWE reported data not only on the frequency of incidents, but also on
the size of the hole in the pipe or the extent of the damage to the pipe in
each case.)

The third pipeline system mentioned above which might on occasion be a
valuable source of statistical data is the high pressure gas transmission
system operated by the British Gas Corporation.   Table 4 gives a summary
of the data used in Reference 1.   In particular for this system data was
available which gave the frequency of occurrence against size of damage as
follows:

Equivalent hole diameter greater than 80 mm   1 incident
    "   "   "   20 to 80 mm 3 incidents
    "   "   "   less than 20 mm 27   "

Thus far, therefore, it is suggested that an acceptable pipeline hazard assessment
will have identified a suitable statistical basis for the assessment, will have
refined the statistics as appropriate to describe the position in respect of
the pipeline in question, and will have expressed the statistics to show the
relationship between frequency of occurrence and pipeline damage hole size.

4. FREQUENCY OF INTERACTION OF MATERIAL FROM THE PIPELINE WITH NEARBY POPULATION

This Section deals with estimation of the number of occasions when, following a
leakage from a pipeline, material in concentration above the lower flammable or
toxic level may be expected to reach a nearby population.

Developing on the ideas described so far. a relationship between expected
frequency of occurrence per kilometre per year and the damage hole size will
be available.   Estimated rates of leakage for each important hole size,
depending on the physical form of the escaping material, whether gas, liquid
or flashing aerosol, should then be calculated by established methods, if
possible taking account of jet entrainment dilution effects.   These rates
should be used to predict the distance to the lower flammable or toxic limit,
as appropriate, by means of suitable techniques for describing atmospheric
dispersion.

The population — housing, schools, hospitals etc—most exposed to the pipeline
by virtue of distance and topography should be identified, and the length over
which each is parallel to the pipeline and the distance of this length from the
pipeline estimated.   Schematically the situation may be described as in Fig 1,
where P is the length of population parallel to the pipeline and D is the
distance between the population : and the pipeline.   D is equally the minimum
distance of travel necessary at any rate of release for the population to be
exposed to a particular level of concentration, which for flammables is, say,
the LFL.   Thus for each of the rates of release calculated for the given hole
size, called R_1, R_2 and R_3 in ascending order, it is possible to determine
the dispersal distances to the LFL: D_{1A}, D_{1B} etc, D_{2A}, D_{2B} etc and D_{3A}, D_{3B} etc
for each weather category A to G.

From these dispersal distances it is possible to obtain the length of pipeline
that can interact with the population as P_{1A}, P_{1B} etc P_{2A}, P_{2B} etc and P_{3A},
P_{3B} etc.   By multiplying these by the frequency that the release might occur,
F_1, F_2 and F_3, it is possible to calculate the expected frequency that a
particular release rate will affect the population under a given weather
category, ie f_{1A}, f_{1B} etc, f_{2A}, f_{2B} etc and f_{3A}, f_{3B} (where $f_{1A} = F_1 \times P_{1A}$).

These frequencies can be modified by two other factors.   Firstly there is
the direction or wind factor, which is the probability that the material will
go in the direction of the population to give frequencies f'_{1A}, f'_{1B} etc,
f'_{2A}, f'_{2B} etc, f'_{3A}, f'_{3B} etc (where $f'_{1A} = f_{1A} \times w_f$ etc.)   This factor can
be applied in most cases as the frequency that the wind is in the required
direction, but in special circumstances, as for example when the escaping
vapours are denser than air and the populations are downhill from the pipeline,
the factor might be taken at 1.0.

The second factor is the frequency of occurrence of the various weather
categories, WF, to give rise to frequencies f''_{1A}, f''_{1B} etc, f''_{2A}, f''_{2B} etc.
f''_{3A}, f''_{3B} etc (where $f''_{1A} = f'_{1A} \times W_f = F_1 \times P_{1A} \times W_f \times WF$).

The total frequency that the pipeline might affect a given population at a given release rate is therefore the sum of the frequencies relating to that rate, thus:

Interaction Frequency Release Rate 1 $= f''_1 = f''_{1A} + f''_{1B} + f''_{1C} +$ etc

and the total for all conditions is:

Total Interaction Frequency $= f''_1 + f''_2 + f''_3$.

These various steps are shown in the brief description of the application of this approach to pipeline safety evaluation which is given in Section 5 of this paper.

Also, as mentioned in Section 2.4, the effects of a leakage can be extended to estimate the consequences of the ignition of the flammable vapours. Thus, strictly, the risk to the public can only be fully assessed if the consequences of interaction of material from a pipeline with nearby population in terms of expected injuries to people and damage to property are known. It is apparent that the frequency of injury or damage will not be greater than the interaction frequency (outlined above) as the probability of ignition and the subsequent probability of injury are less than 1.0. In a particular application it may be possible to show that sufficient precautions, such as increased pipeline wall thickness, depth of soil cover and the monitoring system, are such that the interaction frequency represents the maximum frequency of injury and damage, and if this is acceptable, it is unnecessary to carry the assessment further.

5. APPLICATION OF THE APPROACH

The safety assessment given in Reference 1 was based on the approach outlined in this paper and, by way of illustration, the main aspects are described below. The hazard was the flammability of the natural gas liquids whose approximate composition is shown in Table 5.

5.1 Pipeline Failure Statistics Used

Data from the US system was examined but not used because insufficient time was available to identify pipelines appropriate for comparison with the proposed pipeline.

Statistics from the European petroleum network were used after elimination of inapplicable categories of failure.

$$\underline{\text{Identified failure rate } 2.31 \times 10^{-4} \text{ km}^{-1} \text{ yr}^{-1}}$$

The European statistics were compared with statistics from the British Gas high pressure system after elimination of the same categories of failure.

$$\underline{\text{Identical BG failure rate } 2.32 \times 10^{-4} \text{ km}^{-1} \text{ yr}^{-1}}$$

British Gas supplied data on the distribution of failure rate with damage hole size as shown in Table 6.

Hole Diameter (mm)	Distribution of Incidents	Estimated Frequency km^{-1} yr^{-1}
> 80	1	7.5×10^{-6}
20–80	3	2.2×10^{-5}
< 20	27	2.02×10^{-4}
Total		2.32×10^{-4}

This distribution of failure rate versus hole size was assumed for the natural gas liquids pipeline.   The rate for holes greater than 80mm diameter was regarded as too imprecise and was not used further.

5.2   Atmospheric Dispersion of Vapours from Natural Gas Liquids

Estimated leak rates: 20mm hole — 15.2 Kg s^{-1}

80mm hole — 100 Kg s^{-1}

The effects of gravity slumping in possibly increasing the distance to the lower flammability limit, and jet dilution in decreasing the distance, were not estimated.   Distance to the lower flammability limit was calculated using Bryant's method (Reference 3). (Table 7)

Table 7

Distances to the Lower Flammability Limit

Leak Rate $kg\ s^{-1}$	Expected Frequency km^{-1} yr^{-1}	Distance at LFL — m		
		Cat D	Cat E	Cat F & G
15.2 (R_1)	2.0×10^{-4} (F_1)	105 (D_{1D})	200 (D_{1E})	400 (D_{1F})
100 (R_2)	2.2×10^{-5} (F_2)	320 (D_{2D})	600 (D_{2E})	1300 (D_{2F})

Distances in A, B and C weather were too small to reach significant populations.

5.3   Frequency of Interaction with Nearby Populations

The assessment dealt with eleven populations near the pipeline.   These may be illustrated by Scotlandwell (S) (wind direction effects in atmospheric dispersion) and Glenfarg (G) (gravity slumping effects). (Table 8)

Table 8

D Distance	P. Length	Interaction Length					
		Cat D		Cat E		Cat F & G	
		(R_1) 15.2	(R_2) 100	(R_1) 15.2	(R_2) 100	(R_1) 15.2	(R_2) 100
S 150	150	0	600	400	1250	800	2500
		(P_{1D})	(P_{2D})	(P_{1E})	(P_{2E})	(P_{1F})	(P_{2F})
G 250	500	0	1200	0	1750	1200	3000

Multiplication by the frequency of occurrence of each leak size by the
interaction length of pipeline (P_{1D} etc) gave the provisional expected frequency
of interaction. (Table 9)

Table 9

Location	Provisional expected frequency of interaction yr^{-1}					
	Cat D		Cat E		Cat F & G	
	R_1 15.2	R_2 100	R_1 15.2	R_2 100	R_1 15.2	R_2 100
S	0	1.3×10^{-5}	0.8×10^{-4}	2.8×10^{-5}	1.6×10^{-4}	5.5×10^{-5}
	(f_{1D})	(f_{2D})	(f_{1E})	(f_{2E})	(f_{1F})	(f_{2F})
G	0	2.6×10^{-5}	0	3.9×10^{-5}	2.4×10^{-4}	6.6×10^{-5}

Wind would blow from the pipeline to Scotlandwell for 39% of the time
approximately, at Glenfarg gravity slumping would dominate. Thus, the
frequency of interaction if modified, Scotlandwell by 0.39, Glenfarg by 1.0.
(Table 10)

Table 10

Location	Provisional expected frequency of interaction yr^{-1}					
	Cat D		Cat E		Cat F & G	
	R_1 15.2	R_2 100	R_1 15.2	R_2 100	R_1 15.2	R_2 100
S	0	5.1×10^{-6}	3.1×10^{-5}	1.1×10^{-5}	6.2×10^{-5}	2.1×10^{-5}
	(f'_{1D})	(f'_{2D})	(f'_{1E})	(f'_{2E})	(f'_{1F})	(f'_{2F})
G	0	2.6×10^{-5}	0	3.9×10^{-5}	2.4×10^{-4}	6.6×10^{-5}

In the geographical area of interest, categories D, E and F & G occur 65%, 6% and 8% of the time respectively. Thus, the expected frequencies of interaction were again modified. (Table 11)

Table 11

Location	Frequency of Interaction yr^{-1}					
	Cat D		Cat E		Cat F & G	
	R_1 15.2	R_2 100	R_1 15.2	R_2 100	R_1 15.2	R_2 100
S	0 (f''_{1D})	3.3×10^{-6} (f''_{2D})	1.9×10^{-6} (f''_{1E})	6.6×10^{-7} (f''_{2E})	5.0×10^{-6} (f''_{1F})	1.7×10^{-6} (f''_{2F})
G	0	1.7×10^{-5}	0	2.3×10^{-6}	1.9×10^{-5}	5.3×10^{-6}

The total expected frequencies of interaction for Scotlandwell and Glenfarg were then obtained by summarising the contributions from releases in the 15.2 kgs^{-1} and 100 kgs^{-1}. (Table 12)

Table 12

Total Interaction Frequencies — Incidents per year

	R_1 15.2 kgs^{-1}	R_2 100 kgs^{-1}	Total
S	6.9×10^{-6}	5.6×10^{-6}	1.3×10^{-5}
G	1.9×10^{-5}	2.5×10^{-5}	4.4×10^{-5}

The safety evaluation of Reference 1 showed Scotlandwell and Glenfarg to have interaction frequencies relatively higher than other populations along the proposed route. Subsequent re-routing proposed at these places has brought the total interaction frequency to 1.4×10^{-6} for Scotlandwell and to 1.7×10^{-6} for Glenfarg, which compares with the other calculated frequencies.

REFERENCES

1.   Health & Safety Executive   "A Safety Evaluation of the Proposed St Fergus
to Moss Morran Natural Gas Liquids and St Fergus to Boddam Gas Pipelines"
(July 1978)

2.   National Transportation Safety Board   "Special Study – Safe Service Life
for Liquid Petroleum Pipelines" (Report No NTSB–PSS–78–1 October 12 1978)"

3.   Bryant P M   "Methods of Estimation of the Dispersion of Windborne Material
and Data to Assist in their Application"   United Kingdom Atomic Energy
Authority Report AHSB(RP)R42, 1964.

ACKNOWLEDGMENTS

The Authors wish to express their thanks to the Health and Safety Executive
for permission to publish this paper.

<u>TABLE 1</u>

<u>Recent Significant Pipeline Incidents</u>
<u>Flammable Material Released and Ignited</u>

<u>Location</u>

	Franklin County (Port Hudson) Missouri	Austin, Texas	Meridian, Mississippi	Devers, Texas	Romulus, Michigan	Ruff Creek, Pennsylvania
Date	9.12.70	22.2.73	21.5.74	12.5.75	2.8.75	20.7.77
Material	propane	natural gas liquids (methane to hexanes)	natural gas	ethane-propane mixture	propane	propane
Pipeline diameter (m)	0.2	0.25	0.15	0.2	0.2	0.3
Cause of Failure	corrosion	pipe stress	internal corrosion and em-brittlement	pipe stress	crack propuga-tion	stress-corrosion cracking
Nature of Incident	explosion	fire	fire	explosion and fire	fire	fire
Fatalities	0	6	5	4	0	2
Ref NTSB No.	PAR-72-1	PAR-73-4	PAR-76-1	PAR-76-5	PAR-76-7	PAR-78-1

TABLE 2

Recent Experience on United States Pipelines*

A. Reported Accidents. All Causes - All products

Year	Reported number of Accidents	Reported Number of Trunk Pipeline Miles	Accident/Leak Rate Per Mile
1968	421	115,238	.0037
1969	350	117,983	.0030
1970	288	122,365	.0024
1971	258	122,471	.0021
1972	235	124,458	.0019
1973	194	122,354	.0016
1974	199	126,211	.0016
1975	180	121,278	.0015
1976	169	//	//

B. Accident Causes. All products 1968-1976

Accident Cause	Number of Occurrences	Percentage
Corrosion	1,024	45%
Equipment Rupturing Line	635	28%
Defective Pipe	165	7%
Defective Welds	125	5%
Incorrect Operations	53	2%
Other	292	13%
Total	2,294	

*Taken from United States National Transportation Safety Board Report No. NTSB-PSS-78-1.

//Data not yet available.

<u>TABLE 3</u>

<u>Data on Oil Company Cross-Country Pipelines in Europe (CONCAWE)</u>

FIVE-YEAR COMPARISON BY CAUSE, VOLUME, EFFECT — 1972-1976

	1972	1973	1974	1975	1976	72-76
COMBINED LENGTH (km x 1000)	15.8	17.3	17.35	17.9	18.1	-
COMBINED THROUGHPUT (m^3 x 10^6)	433	558	524	463	540	-
Number of incidents						
MECHANICAL FAILURE						
Construction	-	2	3	-	2	7
Material	3	5	-	3	3	14
OPERATIONAL ERROR						
System	-	-	-	2	-	2
Human	-	-	-	2	-	2
CORROSION						
External	9	8	7	8	2	34
Internal	-	-	2	-	-	2
NATURAL HAZARD						
Subsidence	-	1	-	-	2	3
Flooding	-	-	-	-	-	-
Other	-	-	-	-	-	-
THIRD PARTY ACTIVITY						
Accidental	8	1	6	5	4	24
Malicious	-	-	-	-	-	-
Incidental	1	3	-	-	1	5
	21	20	18	20	14	93
SPILLAGE PER INCIDENT						
Negligible	5	4	6	3	2	28
In excess of 1 m^3	4	3	5	5	1	19
In excess of 10 m^3	6	9	3	5	6	32
In excess of 100 m^3	6	4	4	1	4	19
In excess of 1000 m^3	-	-	-	-	1	1

Taken from Table A 7-2 of Reference 1.

TABLE 4

Data on UK gas transmission pipeline (January '69 to December '77)

FAULTS AND FAULT RATES ASSOCIATED WITH LOSS OF GAS

	Pipeline Diameter mm (Inches)							
	less than 460 (18)	460 (18)	508 (20)	610 (24)	760 (30)	915 (36)	1,070 (42)	TOTAL
System length km	5,769	1,958	277	2,927	798	2,485	34	14,228
Experience km – years	68,300	18,996	2,839	25,995	5,805	11,418	323	133,676
Fault Cause	Number of Faults and Faults per 1,000 km – year							
External Interference	15	4	0	1	1	0	0	21
	0.22	0.21	0	0.04	0.17	0	0	0.16#
Corrosion	18	2	0	1	0	0	0	21
	0.26	0.10	0	0.04	0	0	0	0.16#
Material or Construction Defect	5	0	0	3	1	0	0	9
	0.07	0	0	0.12	0.17	0	0	0.07#
Ground Movement	1	0	0	0	0	0	0	1
	0.01	0	0	0	0	0	0	0.007#
Other	8	7	1	7	0	0	0	23
	0.12	0.57	0.35	0.27	0	0	0	0.27#
TOTAL	47	13	1	12	2	0	0	75
	0.68	0.68	0.35	0.46	0.34	0	0	0.56#

#Fault Rate based on total pipeline system length of 133676 km – years.

The data in this table is based on information provided by the British Gas Corporation and taken from Table A 7–4 of Reference 1.

Table 5   Fluid composition for typical NGL pipeline operation

(Mole percent)

	Initial operation	Winter operation	Summer operation	Max design operation
Carbon dioxide	0	1.40	1.78	1.68
Methane	0	0.29	0.48	0.20
Ethane	9.62	40.73	49.44	50.74
Propane	59.82	38.37	32.24	31.52
iso-Butane	5.93	3.75	3.13	3.09
n-Butane	14.78	9.24	7.73	7.62
iso-Pentane	2.19	1.35	1.12	1.12
n-Pentane	2.84	1.79	1.50	1.48
n-Hexane	2.01	1.36	1.14	1.12
n-Heptane	1.48	0.89	0.75	0.74
n-Octane	0.78	0.53	0.44	0.44
n-Nonane	0.55	0.30	0.25	0.25
	100.00	100.00	100.00	100.00

(Taken from Table 2 of Reference 1)

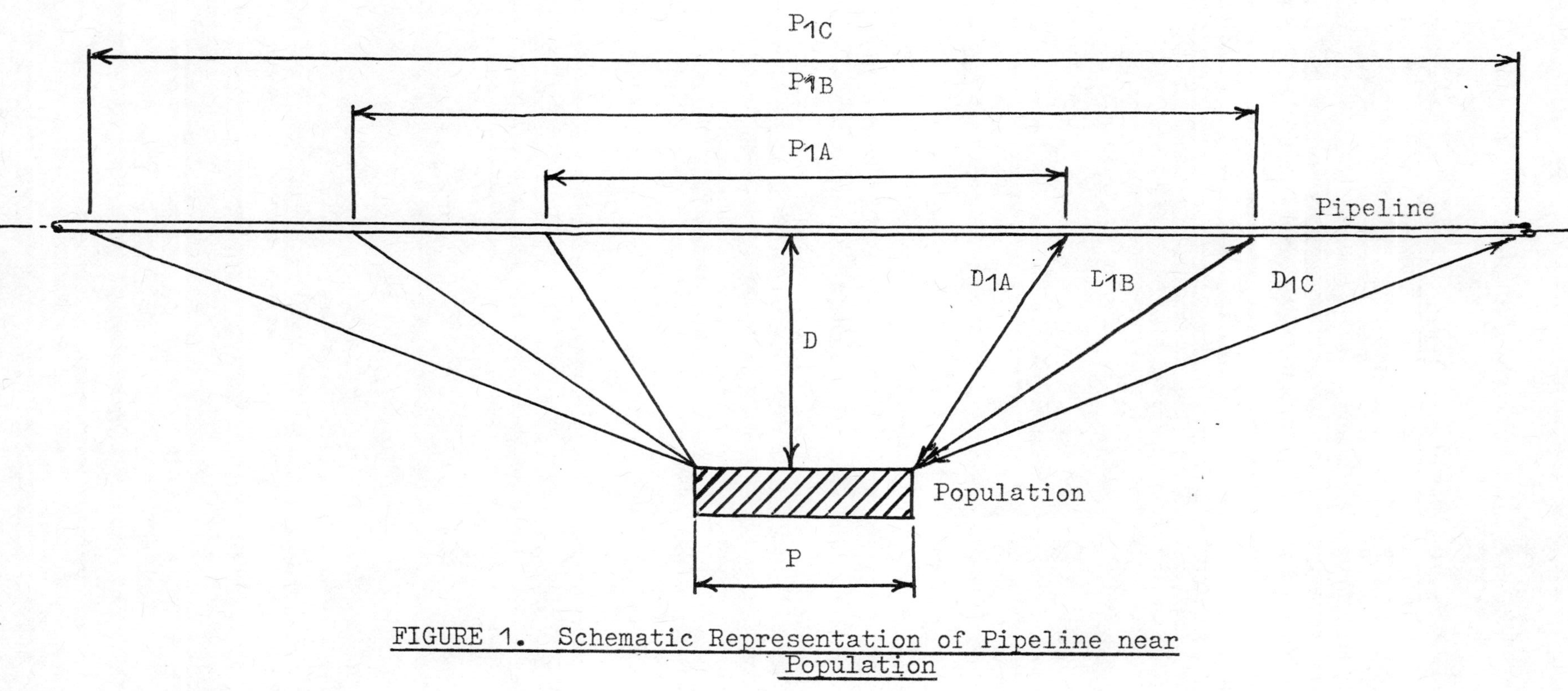

FIGURE 1. Schematic Representation of Pipeline near Population

ınternal and external
protection of pipes

ACCIDENTS WITH PIPELINES IN THE U.S.A.; SELECTED CASE HISTORIES AND A REVIEW OF THE ACTIVITIES OF THE NATIONAL TRANSPORTATION SAFETY BOARD

R.V. Riley, C.Eng., B.Sc., Ph.D., F.I.M., S.C.I.

Ralph V. Riley and Associates, U.K.

Summary

The U.S.A., with over 200,000 km. of Interstate Liquid petroleum pipelines and other pipelines, has a great accumulated experience of pipeline hazards. Since April 1967 all serious accidents with pipes have been surveyed and reported upon by the National Transportation Safety Board which is now a completely independent and autonomous body.

Several incidents involving, Natural Gas Transmission and Distribution mains and pipelines carrying liquid hydrocarbon and chemicals are reviewed. Fourteen case histories are given by way of example; each illustrative of the type of hazard that can occur.

Corrosion as a cause of pipeline failure is now second to damage by human error and external interference to pipeline installations. The "One Call" procedure is discussed which offers a remedy against haphazard development and damage to pipes by contractors equipment.

Finally typical follow-up procedures are discussed. The lessons from the files of the National Transportation Safety Board may be instructive to us in Europe where widespread distribution of energy and chemicals by long distance pipeline is a relatively new experience compared with that in the North American Continent where pipelines have existed for most of this century.

Held at Imperial College, London, England.
Organised and Sponsored by BHRA Fluid Engineering

The North American Continent probably has a greater
accumulated experience of pipelines, conveying gas,
oil and hydrocarbon liquids than any other area in the
world. The U.S.A. in particular has been concerned
with the transmission and distribution of energy
by pipes since the early days of this century. For
example there are over 200,000 km. of interstate
liquid petroleum pipelines.

However until April 1967, any failure of pipelines or
accidents resulting from energy distribution, remained
relatively undocumented, except in incident files of a
local township, gas utility, or pipeline operator.
In that year the National Transportation Safety Board
was set up in Washington D.C.

The N.T.S.B. had its true origin some years earlier when
its duties formed part of the Civil Aeronautics Board
which had been formed to investigate flying accidents.
The American Congress created N.T.S.B. to investigate
accidents and failures in several modes of transportation
i.e. air, rail, highway, marine and pipeline.

Transportation resulted in 49,575 fatalities in 1976
distributed as follows:-

        Total Road fatalities 44,807
        Total Railway fatalities.......... 1,627
        Total Marine fatalities........... 1,741
        Total Aviation fatalities......... 1,318
        Total Pipeline fatalities.........   82

Deaths through pipeline accidents are therefore few but
the nature of some of these accidents and their damage
potential puts them in a class apart from the rest.

The role of N.T.S.B. is investigatory and not regulatory
being totally independent of all interested parties.

THE NATIONAL TRANSPORTATION SAFETY BOARD

Its headquarters are in Washington D.C. 20594, aided by
twelve field offices scattered across the States (1977).
There are about 350 employees but this number is increasing
slowly, particularly in respect of field officers.
The staff actually employed on pipelines is only a small
part of the total but also is growing over recent years.
The group dealing with pipeline problems is known as the
Pipeline Division and the main offices are in Washington,
Los Angeles and Fort Worth, Texas.

The Board's role is investigatory, not regulatory but its
influence on Federal Regulations is very profound as will
be obvious later when we consider certain case histories.
A Federal statute in 1974, The Independent Safety Board
Act, gave the 'Safety Board' its autonomy and divorced
it completely from the U.S. Department of Transportation
(D.O.T.). The new Safety Board (N.T.S.B.) had increased
responsibility for accident investigations and prevention,
in the fields of pipelines and railroads. In brief, the
Board was charged with the duty of investigating all
fatal pipeline accidents and all such non fatal accidents

as caused severe property damage over $100,000.   Since its
inception in 1967 the Board has investigated about 130
pipeline accidents of which about 40 were considered to
be major accidents. These were given detailed reports
whilst the others were reported more briefly. A follow up
action after each incident is to make recommendations
and over 450 such safety recommendations have been sent
to the appropriate Federal regulatory agency, pipeline
operator, municipalities and State authorities and other
interested parties.

Of the total incidents about 70% are concerned with Natural
Gas and of these 70% were in the distribution and 30%
on long haul gas transmission. The 30% non-gas pipeline
accidents usually involve oil or liquified petroleum gas,
the latter being a growing danger and the consequences
are sometimes horrific. Each cubic metre of pipeline
contain 600 times more energy with liquid propane than with
    natural gas. For example a pipeline carrying LPG
ruptured in Missouri nine years ago. The explosion of
the resulting vapour cloud did not kill anybody because it
was in a sparsely populated rural area so the few people
involved were evacuated promptly and in an orderly manner.
The blast when it occurred was registered on a Seismograph
80k.m. away and was calculated to be the equivalent of
50 tonnes of T.N.T!

It is of interest to follow through a few typical incidents,
consider their cause, reflect on the damage sustained
and note the action taken as a result of the survey by
N.T.S.B.

The great losses arising from liquid propane gas accidents
clearly shows the need for more controls. Action was
taken in August 1978 when a start was made on specific
Federal legislation to cover this medium.

1.   ACCIDENTS WITH NATURAL GAS TRANSMISSION PIPELINES

CASE 1. INTERNAL CORROSION OF PIPES AND EMBRITTLEMENT
         OF PIPES

On May 21st 1974 at about 9.45 p.m. 3 persons living near
Meridan, Mississipi became aware of an unusual roaring noise
and left their house to investigate. They drove down an
unpaved road towards the main road and on seeing a huge gas
cloud they hastily returned to evacuate the rest of the
family. The family then drove back to the paved road in
two cars and on turning into the road both engines stalled
because the natural gas in the atmosphere upset the fuel
mixture at the carburettor. (Figure 1) At this moment
a third vehicle coming from the opposite direction entered
the gas cloud and its engine also stalled. Its four
occupants got out and tried to escape on foot. At 10.05 p.m.
the gas ignited, killed one of these on site and in all,
five people were subsequently taken to hospital grievously
burned, causing the deaths of four of them. 40 Ha of
woodland were also destroyed.

This human tragedy was caused by a 150mm diam.natural
gas gathering pipeline rupturing. The cause of the failure
was thinning of the pipewall and embrittlement of the steel
by hydrogen sulphide and carbondioxide internal corrosion.
(Figure 2)

CASE II. HUMAN ERROR AND LACK OF CONSULTATION

Tragedy struck again on a hot summer afternoon near
Cartwright, Louisiana on August 9th 1976 when the driver
of a road grader was busy helping to re-al ign a road.
Unfortunately neither he nor apparently anyone else at the
site was aware of the existence of a 500mm diameter pipeline
carrying natural gas at 30 Bar which was lurking a few
inches under his machine. (Figure 3). The inevitable
occurred, the grader struck & ruptured the steel pipe.
The escaping gas blew out a small crater and within seconds
ignited. The flames engulfed a wide area, killing six
persons, injuring another and doing extensive damage to
property and woodlands. (Figure 4). Two men near the
grader instinctively fled when, they saw the ground erupting
but were burnt to death by the blast. The machine
operator reversed away and was uninjured. The other
casualties were in homes nearby which were burnt to the
ground.

The failure here was entirely human, primarily due to a
lack of communication between the road constructor and the
pipeline operator and the local authority. (Figure 5).
This lack of communication frequently appears in these
case histories and the remedy proposed by N.T.S.B. will be
discussed at length later.

CASE III. FAILURE AT PIPE WELD

A catastrophic disaster in daylight hours is bad enough
but when it occurs at night time it must be the nearest
thing to 'Dantes Inferno'. On March 15th 1974 at about
3.45 a.m. a natural gas transmission pipeline ruptured
sudden ly in a desert area near Farmington, New Mexico.
Three men driving down a road adjacent suddently found
themselves enveloped in a gas cloud with a stalled engine.
On attempting to restart, the sparking of the starter
motor ignited the gas mixture which after an initial
explosion formed a crater 12 metres long x 5 metres wide and
3 metres deep and burned with a flame 30 metres high. (Figure
6).

The flame burned until 5.00 a.m. when emergency crews
     located and closed two valves, isolating
a 15 k.m. section of the failed pipeline.

The cause of the pipe rupture was the brittle fracture
of a longtitudinal flash weld that had been weakened
by localised crevice corrosion and due partly to the
presence of oxides in the weld. (Figure 7). Gas at
a pipeline pressure of 33 Bar escaped from the 300mm
diameter pipe and fuelled the conflagration. This
was an old pipeline made by a manufacturing process which
has now been superseded by the ElectricalResistance Weld
or the Submerged Arc process of welding. It had
suffered from gas leaks several times previously since
1964 but these had never proved catastrophic before.
The pipeline was uncoated and had suffered from corrosion
on the underside. It had been protected, rather inadequately,
by Cathodic protection. It was recommended that
more Cathodic protection must be applied by 30 amperes
rectifiers not more than 15^{km}. apart.

<u>CASE IV.   FAILURE OF PIPELINE UNDER A ROADWAY</u>

Wherever a pipeline traverses a public highway or railway
it is customary to insert it within a casing pipe
previously thrust bored below the foundations of the road.
In this way the minimum disruption to the service is
caused. This common method of construction resulted in a
major accident near Munroe Louisiana on 2nd March 1974
but fortunately without casualties.   (Figure 8). A 750mm
diameter natural gas line failed at a girdle weld within
the cased section below State Highway 124; a fire resulted which
burned 40 ha of forest and disrupted road traffic for many
hours.
In the month prior to this incident there had been 300 mm of heavy
rain causing flooding and the failure was attributed to pipe movement
inside the casing accentuating the bending stresses of
clay-heave at each end of the cased section. This stressing
combined with a sub standard weld resulted in the failure.

2.   <u>ACCIDENTS AT PUMPING STATIONS</u>

<u>CASE V.   EXPLOSION AND FIRE AT PUMP STATION 8, ALYESKA PIPELINE</u>

One man was killed and 5 injured when a pump station 27 k.m.
S.E. of Fairbanks, Alaska exploded and burst on July 8th
1977. The entire 1200mm diameter crude oil pipeline from
Prudhoe Bay to Valdez was shut down, resulting in $35
million damages. (Figure 9).

After $3\frac{1}{2}$ years of construction the pipeline began pumping
at Prudhoe Bay on June 20th 1977 and as the oil reached
each pump station, it was to start up in sequence. In
the morning of July 7th the scraper at the head end of the
crude oil arrived at pump station No. 8. Nitrogen from
the station pipes and control lines was bled off and the
valves were set to commence pumping.

By 2.26 p.m. on the following day problems began to occur
which stemmed from a blocked fine mesh strainer located
in the suction piping of pump no. 1. A maintenance crew
started work on the strainer but because of a poorly
co-ordinated action the control room personnel started
the pump and oil sprayed out under a pressure of 16 Bar
deluging the whole pump hall, exploding and catching fire.

The enquiry criticised the design of the station which
prevented the pumps being seen from the control room and
the lack of controls in the pump hall to close or overide
the valves. The N.T.S.B. report clearly outlines the
need for more careful managerial definitions of jobs
and a detailed prior training programme.

<u>CASE VI. CRUDE OIL TERMINAL FIRE AT LIMA, OHIO</u>

On January 17th 1975 a motor operated valve downstream
of the terminal closed inadvertently causing 'liquid hammer'
on the delivery side of a pump. Although the system was
designed to handle such a pressure build up, estimated
at 50 Bar, a substandard pipe flange failed. A crack
0.5 metres long quickly developed. Crude oil sprayed out,
atomised and ignited and burnt with a flame 30 metres
high.

The flames melted a H.T. overhead electric cable (Figure 10)

cutting off all power to the terminal and because of this
various safety procedures could not be put into practice.
A delivery man in a truck at the terminal started his
engine in an effort to get clear and this evidently ignited
the oil cloud. The terminal was completely destroyed killing
the driver of the truck.

CASE VII.   EXPLOSION AND FIRE AT A NATURAL GAS COMPRESSOR STATION

At 11.00 a.m. on December 7th 1976, near Robstown, Texas,
a natural gas compressor working at 67 Bar failed causing
an explosion and fire. A worker actuated the automatic
emergency shut down system but four gate valves in two
pipe lines failed to close. Another emergency system failed
to shut off two of the five compressors. At 11.30 a.m.
a second more serious explosion occurred and a fire burned
for 3 hours, killing one person and totally destroying
three compressors causing total damages to property and
loss of gas to the value of $5 million.

The original cause was the failure of overtightened studs
securing a valve cover. (Figure 11). The secondary fire
and huge loss of gas were believed to be due to failure of a
cast iron   compressor cylinder caused by overheating
and the heat prevented several emergency circuits from
working properly.

3.   ACCIDENTS CAUSED BY GAS EXPLOSIONS IN BUILT UP AREAS

Although the cases already studied were horrific in their
scale and magnitude, since they occurred generally in
rural areas or even desert ortundrathrough which transmission
lines usually pass, the loss in life and injuries sustained
were not as great as might be expected. However there have
been several incidents in which relatively small amounts of
 combustibles have contributed to much damage to life and
limb. This is frequently the case when failure to a
gas distribution system happens in a town. One such case
occurred at FremontNebraska on January 10th 1976.

CASE VIII.   EXPLOSION AND FIRE AT PATHFINDER HOTEL, FREMONT, N.A.

The Pathfinder Hotel was a six storey brick built hotel
constructed in 1917, very old by American standards. On
Saturday 10th January 1976 a resident smelled gas. He
notified the night porter at 4.50 a.m. who 'checked' the
second and third floors but claimed he could smell nothing.
The dayclerk arrived at 6.35 a.m. and smelled gas in the
vestibule, went to the kitchen but found nothing wrong.
At 8.02 a.m. a bakery employee from premises across the
street telephoned the hotel saying he had smelt gas in an
alley way near to the hotel. He himself phoned the
Gas Company, (well knowing that he would be responsible
for a $10 service charge for a weekend call-out).
Service men arrived, did not have a gas detector so returned
to the depot. When it was brought to the hotel it did
not work so a man went back again for a replacement.
(Figure 12). By 9- 10, three gas company officials and a
service man arrived. A check made in vaulted basement
recorded 100% methane. The hotel clerk was asked to evacuate
the residents immediately. At 9.32 a.m. it was too late,
the building exploded before the occupants had been warned

to evacuate. The force of the explosion blew up an entire
road surface which went through the bakehouse wall. All
the gas company officials were killed.

A blaze ensued which burned until 2.30 in the afternoon
since no convenient stop tap could be found. In all,
the casualties were 20 persons killed and 30 injured. All
windows were shattered in premises within one block away.
(Fig. 13)
The cause was traced to a gas leakage from a 50mm plastic
pipe which had pulled out of its compression coupling. The
gas permeated the earth and entered basements in the vicinity.
The pipe failure was the result of thermal contraction
over two exceptionally cold winters and an unsuitable coupling.

On December 15th 1977 at Lawrence, Kansas there was a
'carbon copy' incident, also involving a plastic pipe coupling.
This caused the deaths of 2 persons, injured 3 others and caused
extreme property damage.

CASE IX. EXPLOSION AT A 25 STOREY NEW YORK CITY OFFICE BLOCK

At 6.57 a.m. on the morning of 22nd April 1974 a massive low
order explosion demolished the west wall of the Consolidated
Edison building in New York City. The structure of an
adjoining building was damaged and windows were blown out
over a large area. Seventy persons were injured.

The cause was worryingly mundane; a water pipe attached
to a hydropneumatic tank came adrift at a pressure of 12 Bar.
Part of the steel cylinder rocketted upwards, hitting a
steel gas pipe and dragging it from its coupling. The
gas pressure was only 7 ins. w.g. but some 1,500 cubic
metres of gas escaped to form an explosive mixture in the
lift shaft. (Figure 14).

CASE X. APPARTMENT COMPLEX DESTROYED AT EL PASO, TEXAS (Fig.15)

Leakage from a joint in a cast iron reducer from a 150mm cast
iron steel main and two leaks caused by corrosion of an
attached 32mm steel service pipe working at 1.2 Bar pressure
allowed gas to seep under a concrete road. (Figs. 16 & 17).
An explosion resulted in 7 persons being killed, 8 injured
and 7 out of 15 units in the appartment complex were
wrecked. The uncovering and disturbing of the cast iron
pipe 6 days previously by workmen looking for a leak was
a contributory cause to the failure of the reducer gland. A
further cause was the inability of foreign labour on the
maintenance staff to read instructions and notices properly.

Recorded incidents resulting from gas leaks in built up
areas are numerous and it is not possible in the course of this
review to deal with them all. They are frequently the result of
corrosion: in America many steel service lines have in the
past been laid unprotected. Unfortunately the newer polyethylene
gas service pipes, although not suffering from corrosion still
fail, by coupling pull-out or by poor welds particularly to
fittings.

Another frequent cause of gas explosions and fires is
undetected damage to service pipes which allow seepage into
premises and into basements of property sometimes not even
served by gas. Detection of this type of damage is often
difficult and frequently problems in reporting and taking

the necessary action to evacuate premises occur. This sort
of problem has prompted the special attention of the N.T.S.B.

4.   ACCIDENTS CAUSED BY PIPELINE FAILURES WITH CHEMICAL AND
HYDROCARBON LIQUIDS

CASE XI. ANHYDROUS AMMONIA LEAK NEAR CONWAY, KANSAS

Figure 16 shows dramatically a leak of Ammonia gas from a
cross country pipeline of Mid-American Pipeline Systems,
which allowed 2,138 barrels of Anhydrous Ammonia to volatilise
into the air. (Fig. 18). Two persons were hospitalised with
severe burns to eyes, nose, throat and lungs.

An operator started a pump whilst a gate valve further
downstream remained closed; the line pressure rose from the
normal working pressure of 15 Bar to 80 Bar. The pipe failed
at a dent at the site of two gouge marks $\frac{3}{4}$ metre long where
the pipe had been damaged during back filling of the trench
during laying. This diagnosis was possible because the corrosion
protective coal tar wrapping was also damaged at this point.
(Fig. 19)

CASE XII. ACCIDENT WITH PROPANE 200mm DIAMETER PIPELINE ROMULUS, MICHIGAN

This incident was similar to the previous one in that the cause
was a dented and gouged pipe which had been damaged during laying.
The failure occurred when the sudden closure of a valve sent a
pressure surge along the hydrocarbon liquid pipeline. The pipe
wall ripped open with a brittle Sinusoidal failure. (Fig. 20).
The liquid propane sprayed into the air, vapourised and ignited.
Flames 180 metres high engulfed an area 200 metres diameter,
2,400 barrels of propane were burnt up, injuring 9 persons,
destroying 4 houses, 12 vehicles and damaging 3 other houses.

CASE XIII. NATURAL GAS LIQUIDS EXPLOSION AND FIRE. DEVERS, TEXAS

At 12.28 p.m. on May 12th 1975 a 200 mm diameter steel pipeline
working on mixed propane and ethane liquids at 100 Bar cracked
near a 'gouge' mark. The failure resulted from a combination
of cyclic pressure loading, residual pipe wall stresses and thinning
of the pipe by damage from a back hoe during re-instatement
after an excavation for the insertion of a valve.

Almost at the moment of the blow-out a car entered the area along
U.S. Highway 90 and the usual routine followed; i.e. car stalled,
an attempt was made to start it, ignition of the gas cloud
followed. The 4 occupants were burnt to death but they should be
recorded as martyrs to the cause of pipeline accidents since by
igniting the gas cloud at that precise moment the whole area
was saved from a major explosion which would have been a catastrophe
(Fig. 21). As it was the escaping hydrocarbon burned for two
hours, melted telephone and power lines, warped the railway track
and eliminated a complete woodland. All traffic was halted for
over a day on road and rail in both directions. (Fig. 22).

CASE XIV. PROPANE PIPELINE RUPTURE AND FIRE. RUFF CREEK, PENN.

There are many further interesting case histories on the files
of the N.T.S.B. but for this review one must stop somewhere but
preferably not at Case 13 so this Case 14 is the last!
(Fig. 23). Figure 23 shows a valley across which a 300mm diameter
propane pipeline traversed. At 4.30 a.m. (Estimated) January
12th 1978 the pipe fractured and liquid propane escaped under a

pressure of 15 Bar, vapourising to fill the valley with a white
fog. This went undetected until 6.00 a.m. when a pick-up truck
entered the cloud sudden ly. The engine stalled and of course
the frantic efforts with the starter button set off a flash
fire 100 metres wide which swept up and down the valley for
1500 metres burning everything in its path. The two men in the
truck were killed, power and telephone lines vapourised and 57 head
of cattle and 450 bales of hay were lost. The woodland and
meadows were burnt to cinders.

This old pipeline had previously been used between 1944 and 1968
for natural gas at 95 Bar.

Despite extensive re-validation tests prior to the change to
propane (at 22-30 Bar) the presence of stress corrosion cracks
on the outer part of a bend went undetected. Further cyclic
stresses on the weakened pipe wall resulted in the brittle
failure and that allowed all the gas to escape. Statistically
most cases of stress corrosion cracking have been found in
pipes 20-35 years old which have previously carried or continue
to carry natural gas. (Fig. 24 - cracked pipe).

5.   THE IMPORTANCE OF CORROSION AS A CAUSE OF PIPELINE ACCIDENTS

N.T.S.B. has recently analysed all their reports over a 9 year
period to the end of 1976. This has served to illustrate the
accident trends in the liquid pipeline industry. Where pipelines
have been adequately protected by external coating, cathodic
protection and subject to electronic inspection, the failure
rate due to corrosion has dropped, i.e. from 229 in 1968 to
50 in 1976. Because of the significant decline, corrosion is
no longer the main cause of pipeline accidents but is relegated to
second place behind pipeline ruptures caused by operating equipment
such as contractors earth-moving machines, etcetera. One worrying
form of corrosion remains that of stress corrosion, usually
initiated from the inside pipewall by sour gases containing carbon
dioxide and/or sulphides.

6. THE "ONE CALL SYSTEM"

Several of the case histories quoted reveal the problems that can
arise from un-coordinated efforts of individuals who find
themselves at the scene of a pipeline failure. It is clear also
that some of the accidents need never have occurred if various
individuals had persued adequate enquiries before setting to work
to dig up a road. Several of the states in America now operate
a "one call" system or data and information bank from where
help and advice can be obtained by phoning one telephone number.
Since this system started in 1970 in Michigan fewer accidents
of this kind have occurred. There are now 106 "one call" systems
in 40 States.

7.   FOLLOW UP PROCEDURES AFTER PIPELINE ACCIDENTS

Upon the completion of their investigations the N.T.S.B. take steps
to acquaint all concerned with their findings in an effort to
secure remedial action where this is necessary. Personal letters
on green paper are issued to the principals involved, for example:-

Gas Company President or Pipeline Line Operator Head,
Secr. Department of Transportation Washington D.C.,
Department of Housing and Urban Development Washington D.C.,
Building Code Administrator, Plumbing Code Manager,
American Petroleum Institute, City Mayor,
City Public Service Commission, American Gas Association,
State Governors, Director of State Transportation,

American Society of Mechanical Enquiries, Gas Piping Standards Committee, Secr. for Occupational Safety and Health, Materials Transportation Bureau, Water Utilities, Construction Companies, Pipe Manufacturers, Fittings Manufacturers, Plastics Pipe Institute, etcetera.

Each personal letter briefly summarises the details of the accident, gives the findings of the enquiry and makes recommendations, each of which is 'tailored' to the recipient's responsibilties in the matter. Thus to the Manager of a Gas Company involved in a small house explosion the recommendations might read:-

"(1)   Review and revise emergency procedures and employee
        training programme to ensure that houses are entered and
        checked for gas when a leak is suspected.

 (2)   Review and revise service abandonment procedures to ensure
        that abandoned services are cut off from the main and
        capped."

To the Secretary for Occupational Safety and Health, a U.S. Government Department, N.T.S.B.'s green letter, following a burn accident to men repairing a gas main, the recommendation is:-

        "Establish standards for Gas Industry Safety Clothing
        to protect workers repairing leaking gas pipelines where
        ignition of the gas could cause serious burns".

N.T.S.B. also issue for general release a white broadsheet which briefly describes a recent situation and seeks to give a 'helpful message' to a wide readership.

Another type of follow up action is the publication by N.T.S.B. of extracts from the Federal Register, noting such items as testing techniques, specifications of jointing techniques and requirements, definitions of terms reporting instructions and requirements, Corrosion control, design advice, conversion of pipeline to other uses, off-shore requirements, emergency planning and many others. A most recent publication dated February 8th 1979 is the proposed new Federal Standard for Safety for liquified natural gas.

Finally a comprehensive bulletin is published monthly, printed in blue with a blue "Mast head" known as 'Pipeline Safety'. This gives details of notices, amendments and withdrawals, staff vacancies, safety services and conferences, and training courses, News of new standards and codes, general statistics, interpretation of regulations, etcetera.

CONCLUSIONS

No attempt has been made to compare the U.S. investigatory processes on pipeline safety with those of Europe or the U.K. but perhaps this review of the activities of the N.T.S.B. will serve as an impressive reminder that pipelines are to be taken seriously. Most of the U.K. pipelines have been laid within the last 10 years and the pipes are better manufactured and layed and are better protected against corrosion than many of the American cases reviewed here. However we should not remain unmindful of what can happen in the next 10 to 20 years if the lessons of our Uncle Sam go unheeded. Excavation, ground subsidence, plane accidents, human errors, etc will occur - are we happy with <u>our</u> safety precautions?

<u>DEPARTMENT OF TRANSPORTATION'S PIPELINE ACCIDENT REPORTS</u>

<u>REFERENCES.</u>

(1)   Natural Gas gathering pipeline failure, Near
        Meridan, Mississipi May 21st 1974. N.T.S.B. - PAR - 74 - 1.

(2)   United Gas Pipeline Company to "pipeline rupture and
        fire", Cartwright, Louisiana August 9 1976.
        N.T.S.B. - PAR - 77 - 1.

(3)   Southern Union Gas Company. Transmission pipe failure
        Near Farmington, New Mexico. March 15 1974.

(4)   Michigan-Wisconsin Pipeline Company, Near Munroe,
        Lousisiana, March 2 1974.

(5)   Alyeska Pipeline Service Company. Explosion and
        Fire, Pump Stat. 8. Nr. Fairbanks, Alaska.
        July 8 1977. N.T.S.B. - PAR - 78 - 2.

(6)   Mid Valley Pipeline Company. Crude Oil Terminal Fire.
        Near Lima, Ohio. January 17 1975.

(7)   Exxon Gas System. The Natural Gas Explosion and Fire.
        Robstown, Texas. December 7 1976.

(8)   Nebraska Natural Gas Company, Pathfinder Hotel
        Explosion and Fire, Fremont, Nebraska. January 10
        1976. N.T.S.B. - PAR - 76 - 6.

(9)   Kansas Public Service Company Inc. Explosion and Fire
        Lawrence, Kansas, December 15 1977. N.T.S.B. - PAR - 78 -
        K.

(10)   Consolidated Edison Company, Explosion at 305E, 45th St,
        New York City, April 22 1974.

(11)   Southern Miami Gas Company, El Paso, Texas, Natural
        Gas Explosion, April 22 1973.

(12)   Mid American Pipeline System. Anhydrous Ammonia Leak,
        Near Conway, Kansas, December 6 1973.

(13)   Sun Pipeline Company, Rupture of 8" Pipeline, Romulus,
        Michigan, August 2 1975. (PAR-76-7)

(14)   Dow Chemical U.S.A. Natural Gas Liquids Explosion and
        Fire, Near Devers, Texas, May 12 1975.

(15)   Consolidated Gas Supply Corp., Propane Pipeline Rupture
        and Fire, Ruff Creek, Pennsylvania, July 20 1977.

<u>ACKNOWLEDGEMENTS:</u>

        The Author has extracted freely from Accident Reports
        published by the U.S. National Transportation Safety Board,
        to whom acknowledgement is made. In particular to
        Mr. J. B. King, Chairman, who has given permission to the
        Author to use the reports and other literature published
        by the N.T.S.B. for this purpose.

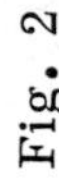
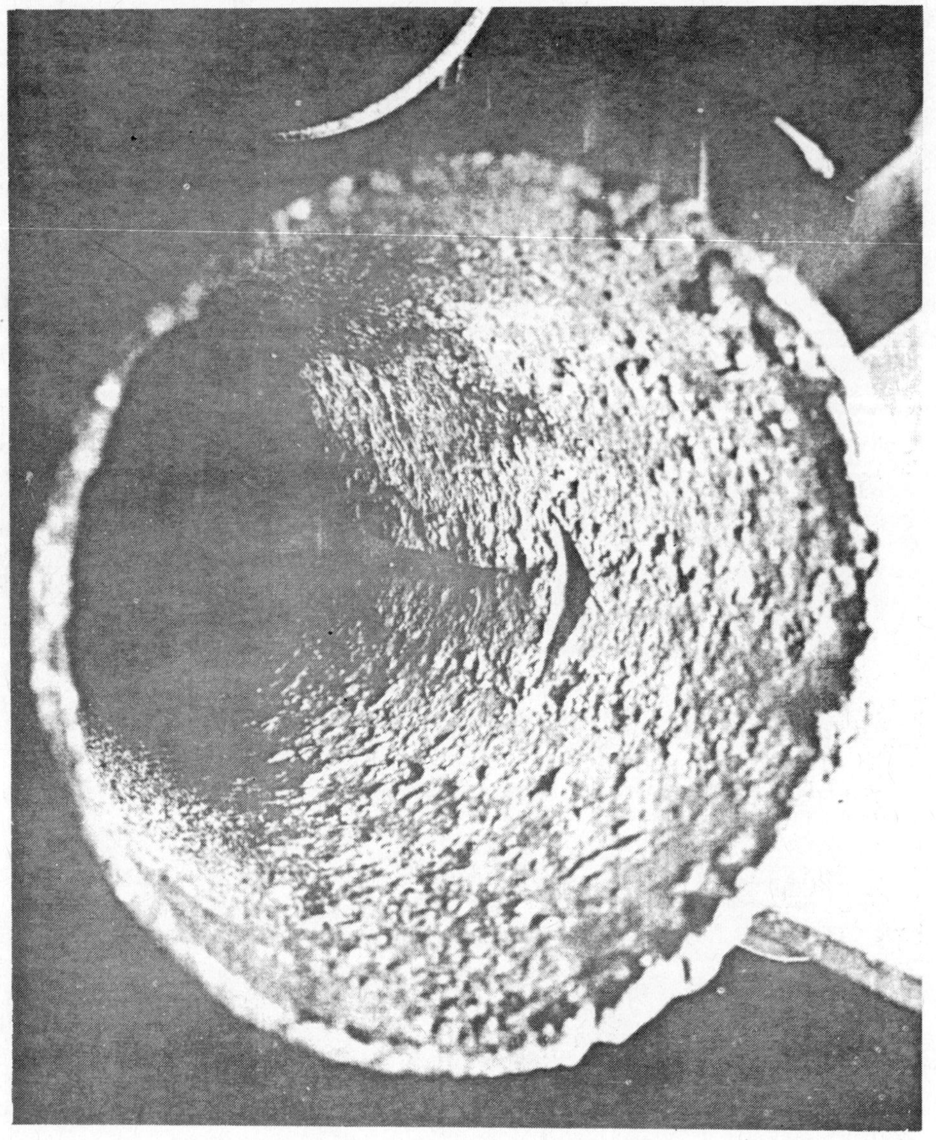

Fig. 2

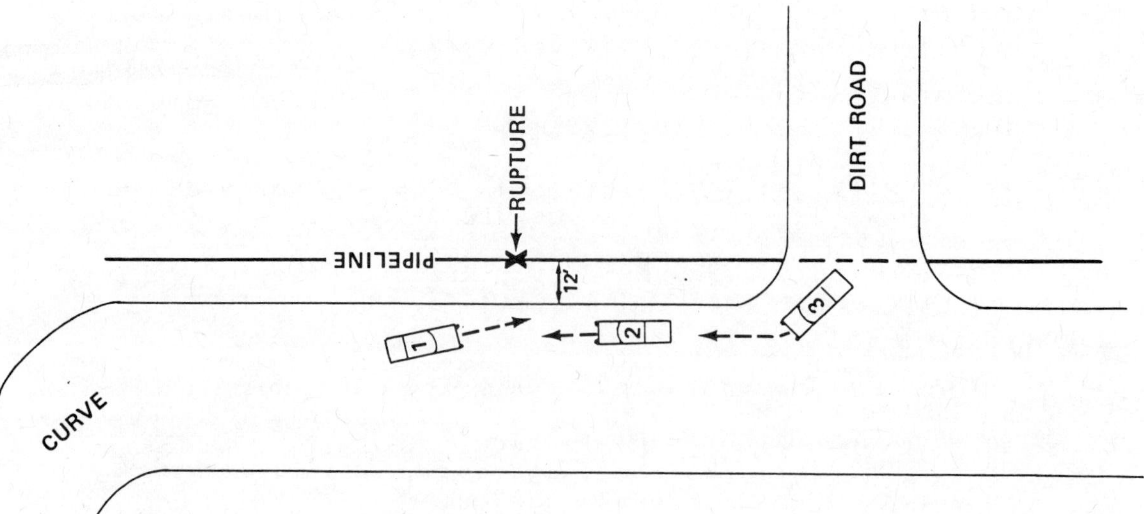

Fig. 1 Accident site.

Fig. 4   Accident site.

Fig. 3   Cross section and profile of east ditch on Griffin Road at 20-inch pipeline crossing.

Fig. 5   Damaged 20-inch pipe.

Fig. 6   Ruptured pipe in the ditch prior to removal. Split was facing down.

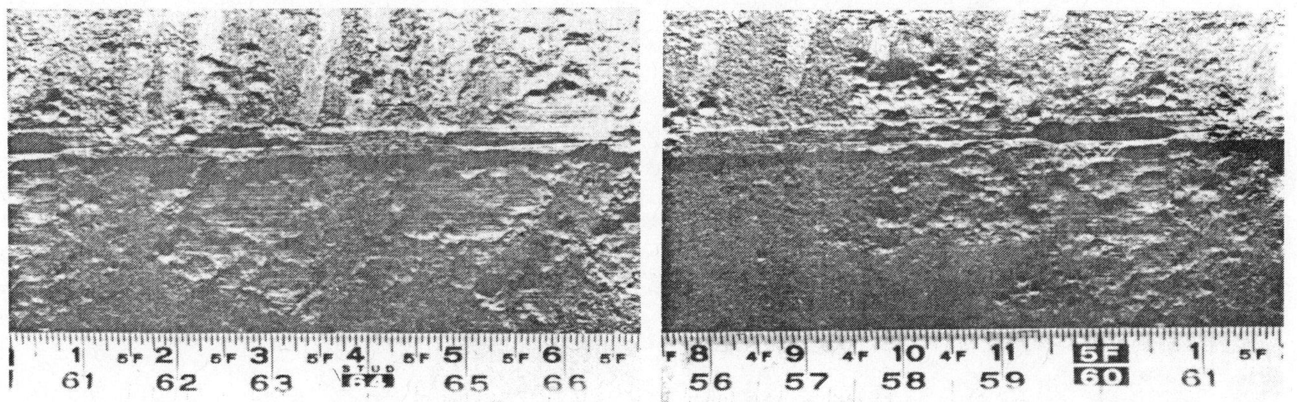

Fig. 7   Close-up views of crevice corrosion in the longitudinal flash weld
and localized attack of adjacent base metal of the north end of the pipe.

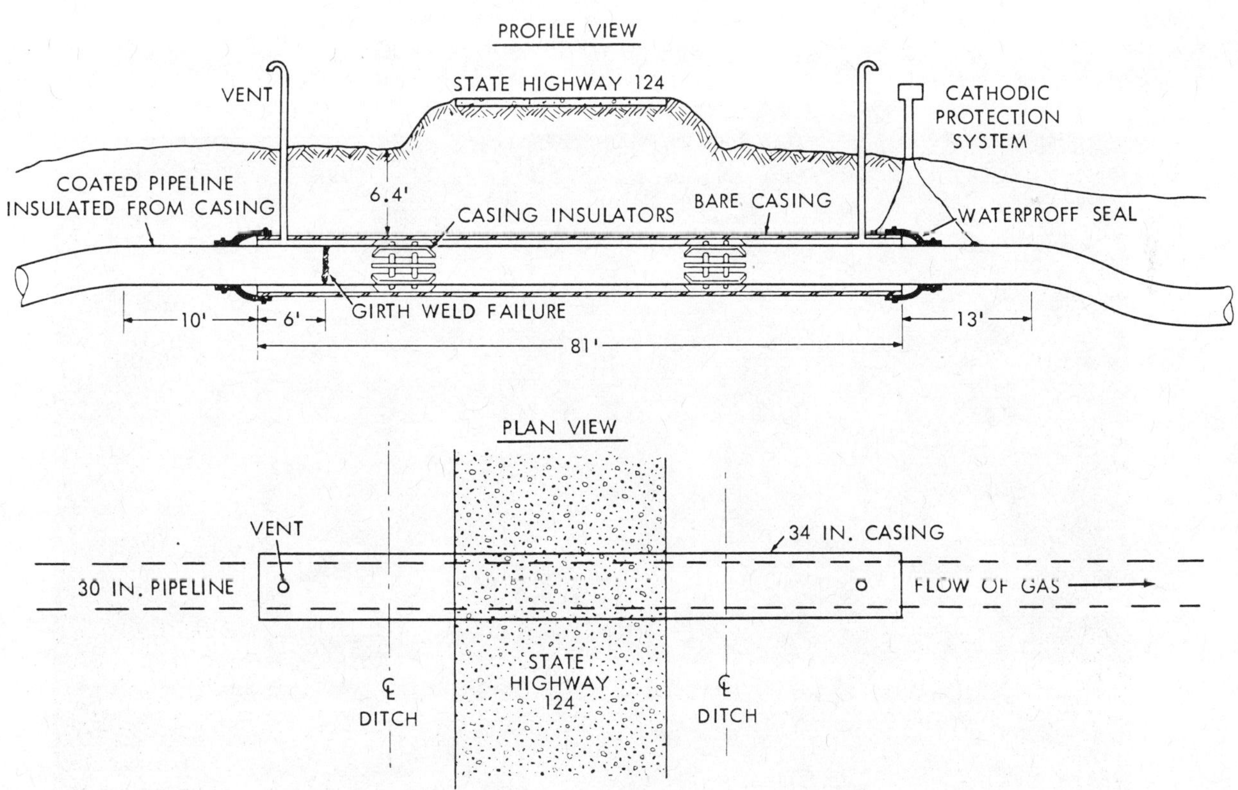

Fig. 8   Pipe in casing under state route 124.

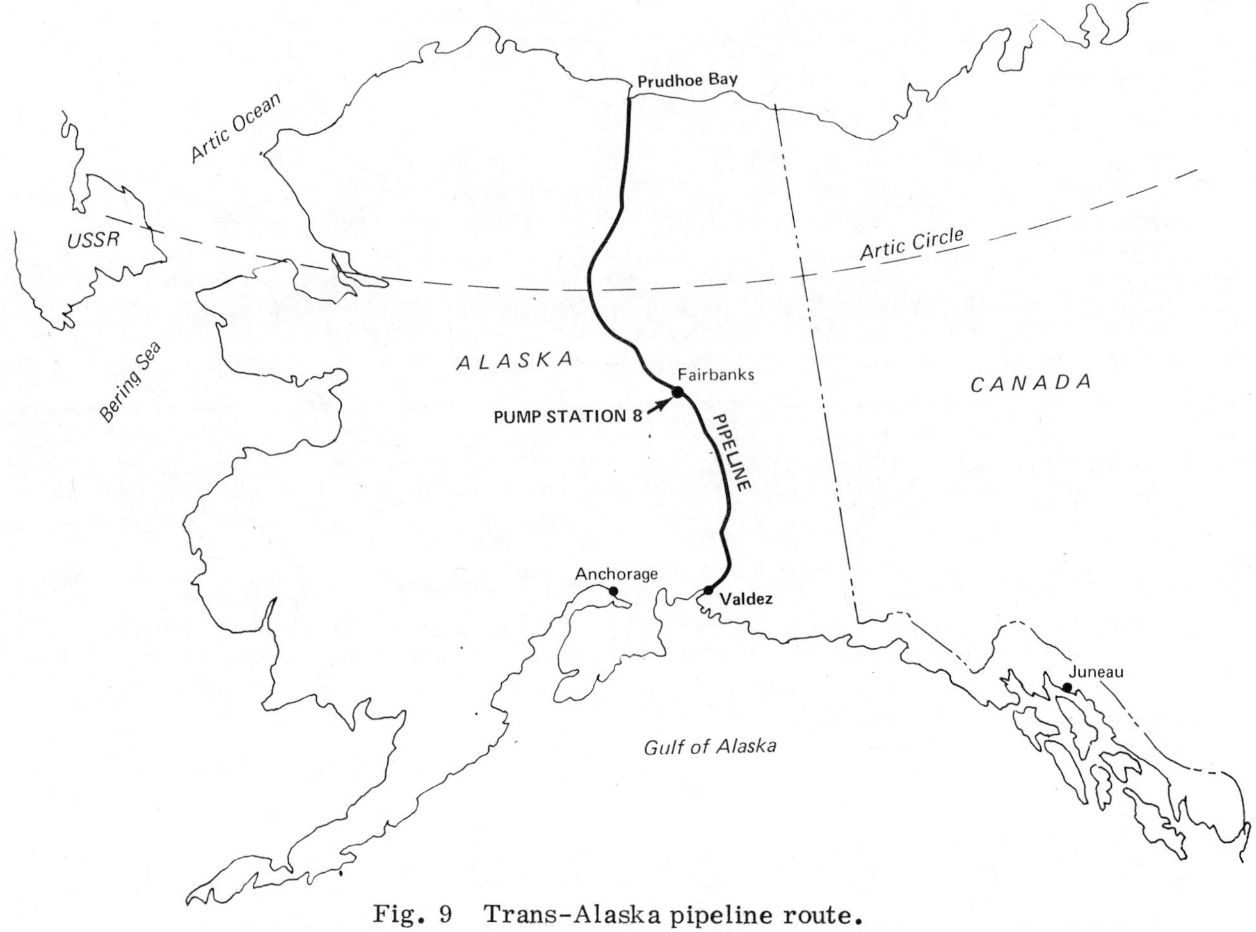

Fig. 9   Trans-Alaska pipeline route.

Fig. 10   Crude oil fire at Lima terminal.

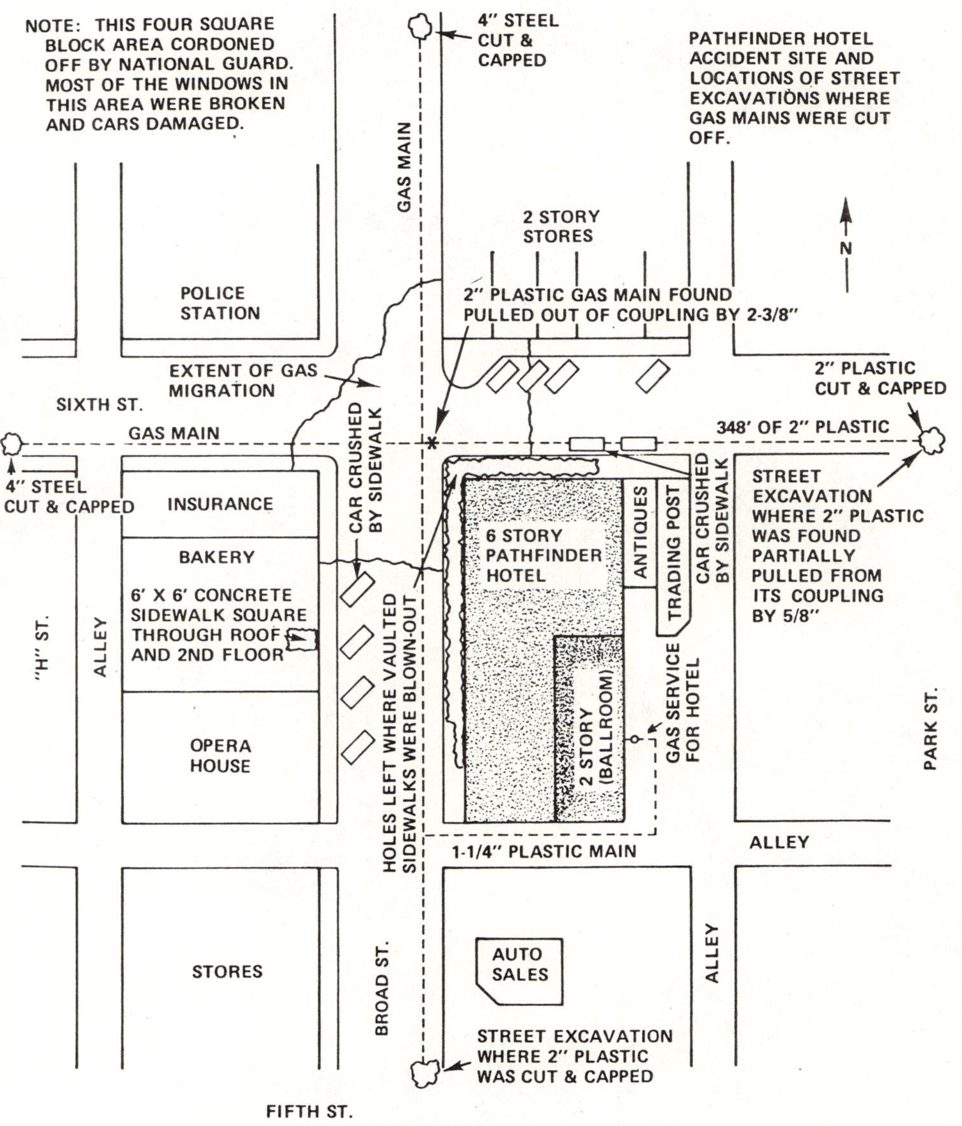

Fig. 11   Failed studs from valve cap No. 3 of cylinder No. 2 of compressor unit No. 1.

Fig. 12   Accident site.

Fig. 13   South and west sides of Pathfinder Hotel.

Fig. 14   Accident site at 305 East 45th Street, New York City.

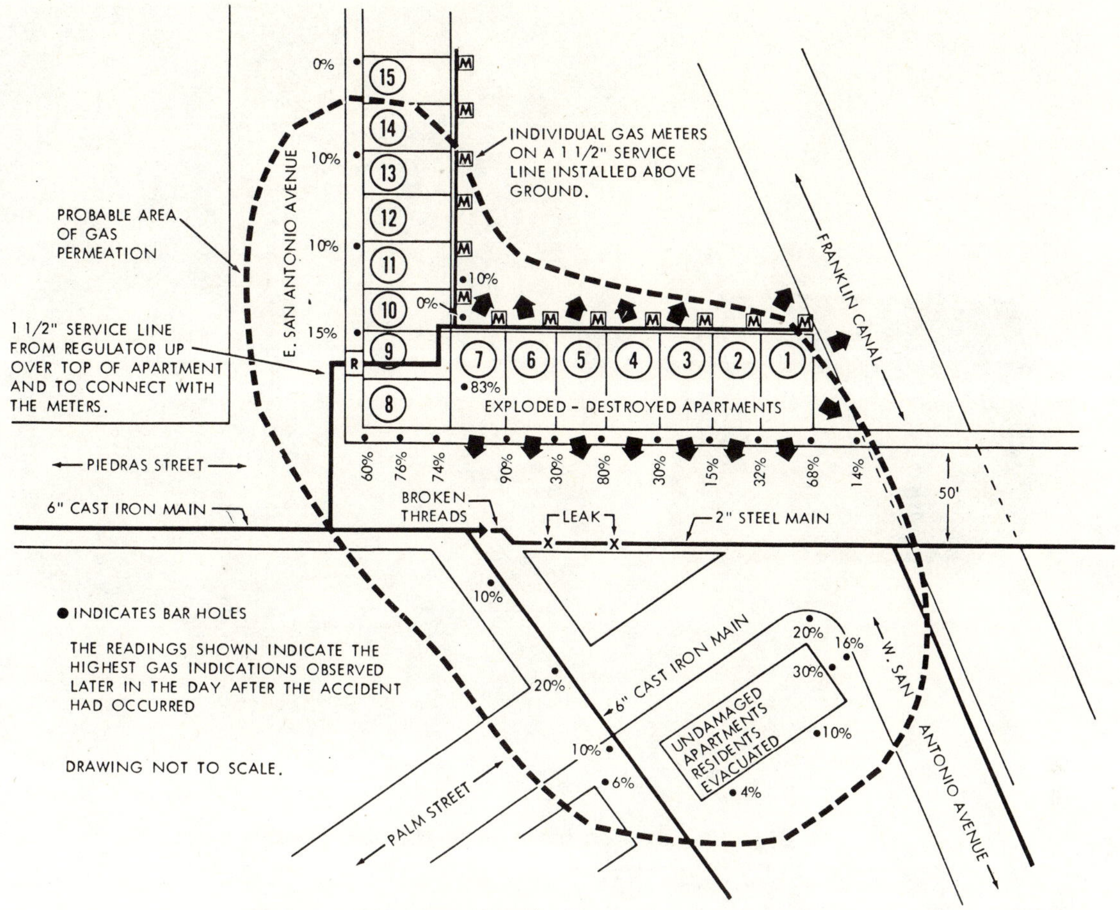

Fig. 15   Accident site.

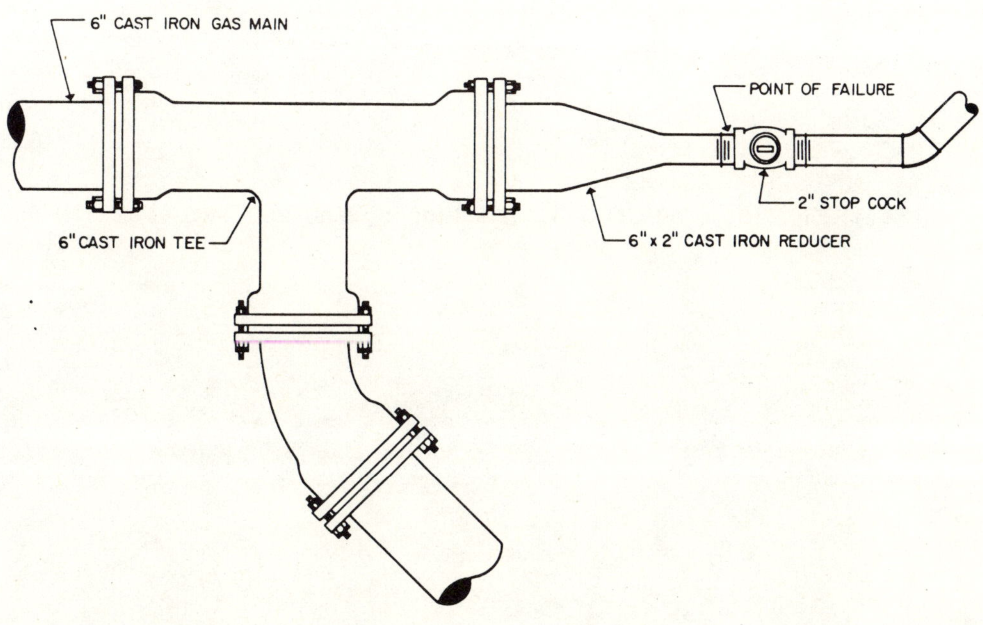

Fig. 16   Diagram of failed pipe section.

Fig. 17   Broken threads at the 2-inch valve.

Fig. 18   MAPCO leak, Hutchinson, Kansas, August 13, 1974.

Fig. 19   Full length of fracture showing its relationship to gouges in the pipe.   The two
horizontal arrows indicate the full length of the gouge.   The two vertical arrows indicate
the extent of the fracture along the gouge. (From report by Metallurgical Consultants Inc.).

202

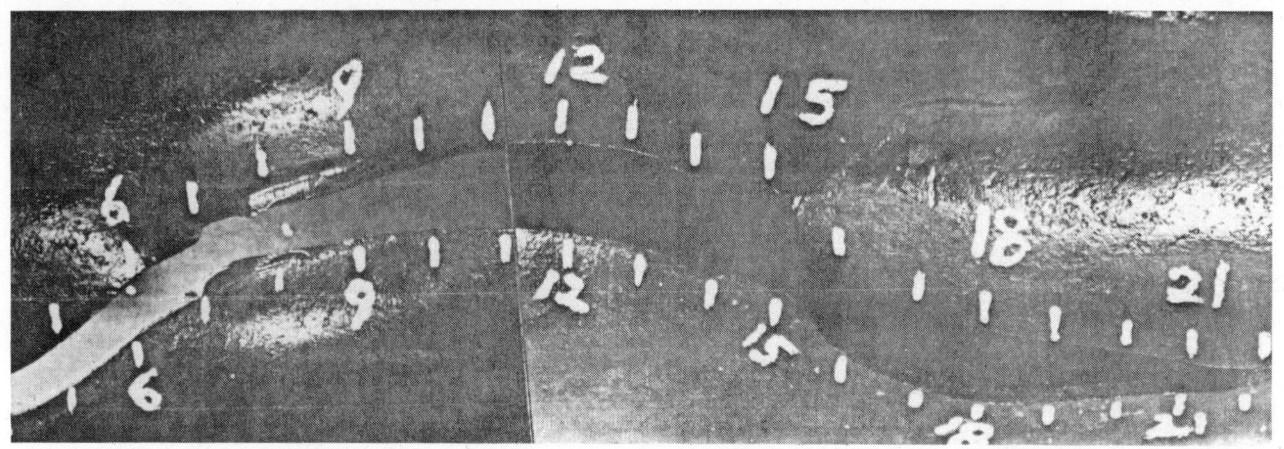

Fig. 20 Composite photograph of the pipe showing the dent, gouge and secondary crack.

Fig. 21 Failed 8-inch pipe.

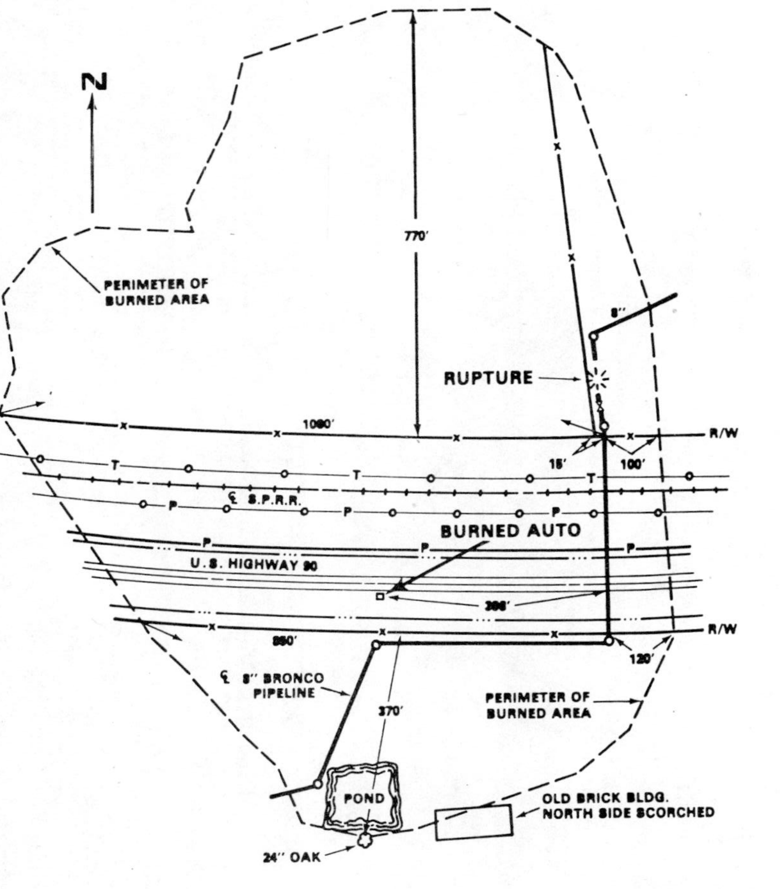

Fig. 22   Accident site.

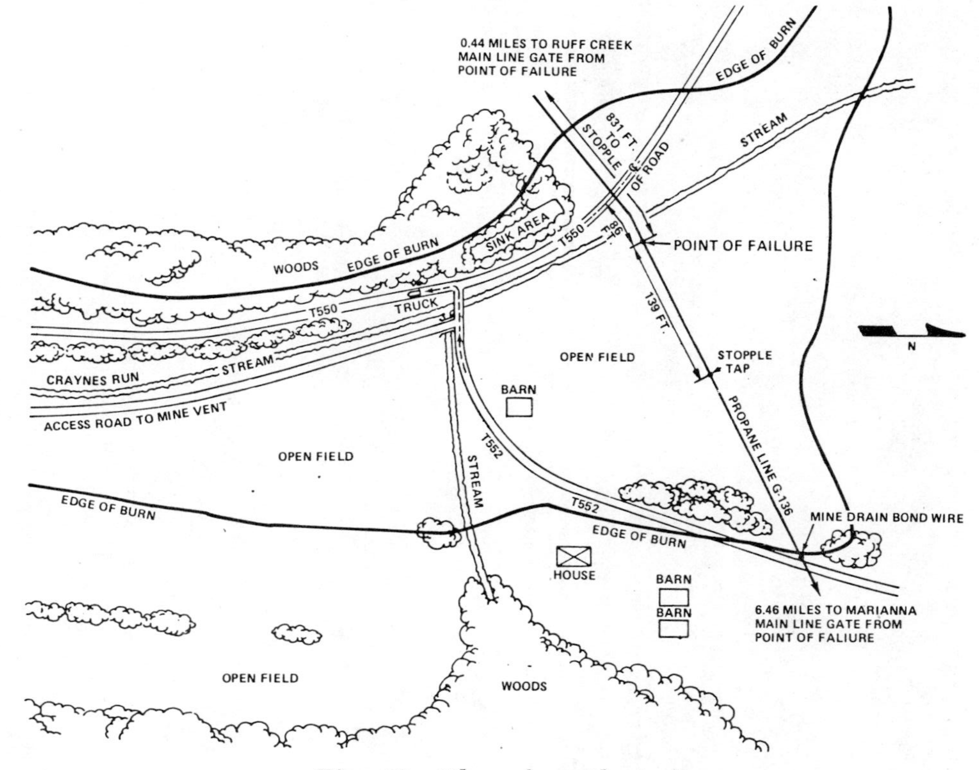

Fig. 23   Plan of accident site.

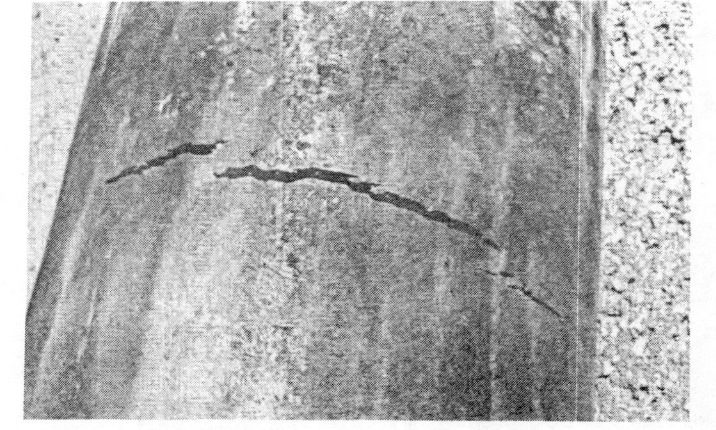

Fig. 24   Ten-inch-long crack in 12 3/4-inch outside diameter, bare steel pipe.

internal and external
protection of pipes

September 5th - 7th, 1979

AQUEOUS CORROSION OF STEEL BY H_2S AND H_2S/CO_2 MIXTURES

D. E. Milliams, PhD, and C. J. Kroese, PhD.

Koninklijke/Shell Laboratorium (Shell Research B. V.), The Netherlands.

Summary

In view of the importance of the corrosion of low-alloy tubular steel for natural gas production and transportation and flooding-operations caused by aqueous solutions containing H_2S or H_2S/CO_2 mixtures, corrosion rate measurements in these media have been carried out.

General corrosion of QMC41 steel has been studied as a function of salt content of the aqueous solution, the temperature and the H_2S and CO_2 partial pressures. A high salt content of the aqueous phase has been found to prevent the build-up of protective surface layers. It also prevents the development of highly anodic surface layer potentials, thus reducing the possibility of pitting corrosion. Introduction of CO_2 in the H_2S-containing corrosive solutions likewise reduces the danger of local attack but only at the cost of higher general corrosion rates.

Held at Imperial College, London, England.
Organised and Sponsored by BHRA Fluid Engineering
Copyright BHRA Fluid Engineering, Cranfield, Bedford, England.

In natural gas production and transportation and also in flooding operations, corrosion of the low-alloy steel pipes by aqueous salt solutions containing H_2S or H_2S/CO_2 mixtures is a phenomenon of rising importance. Apart from the possibilities of stress corrosion cracking and "step-wise cracking" (see e.g. Ref. 1), the corrosion of steel by aqueous solutions containing H_2S or H_2S/CO_2 takes two different forms. The first form is general corrosion, in which the surface is more or less uniformly attacked and an iron sulphide layer is formed which can offer some protection. The second form is pitting corrosion, with high penetration rates.

Although the corrosion of steel by H_2S in aqueous salt solutions had been studied previously (Refs. 2-10), it was felt that a better understanding of the factors involved in the corrosion by H_2S/CO_2 mixtures would first require a preliminary study of the factors governing the corrosion by H_2S alone (partial pressure, temperature and salt concentration). This work is now being extended to corrosion by H_2S/CO_2 mixtures and results of both investigations will be given. Surprisingly, despite its extreme importance in practice, the pitting phenomenon had so far received little attention (refs. 11-14). Several approaches to the simulation of pitting corrosion have been investigated and some preliminary results will be presented.

It is hoped that this study, which is still in progress, will ultimately permit the prediction of corrosion rates in aqueous solutions containing H_2S or H_2S and CO_2. The selection of protective measures and the assessment of their efficiency are also amongst the objectives of this investigation.

EXPERIMENTAL

Most experiments have been carried out in 5.5 l of NaCl solutions of various concentrations to simulate the electrolytes present in pipelines. A gas mixture of appropriate composition was continuously bubbled through the slowly stirred solution, which was kept at constant temperature to maintain fixed concentrations of the dissolved gases.

Weight loss specimens (8.8 cm^2; three in each test) were made of QMC41 Carbon Steel and were usually grit-blasted, degreased, weighed and mounted on a laboratory-made probe. After exposure they were cleaned, dried and reweighed. Corrosion rates were calculated as time averages from the weight losses and the exposure times. The same specimens also served as electrodes for the continuous monitoring of the corrosion rate using a Petrolite Corrosion Rate Meter (automatic model M-1000 series). The readings of the corrosion rate meter were corrected for the influence of the electrolyte resistance between reference and measuring electrodes, which was measured using a Hewlett-Packard 4328 milli-ohm-meter. The readings of the corrosion rate integrated over the full test duration (normally about 120 h) were calibrated by comparison with the weight losses measured.

THE CORROSION OF STEEL BY H_2S-CONTAINING NaCl SOLUTIONS

Some characteristic graphs of corrosion rate readings against time are shown in Fig. 1. In all instances readings are initially high, and drop in the course of time to substantially lower values. This is accompanied by the build-up of more or less protective corrosion product layers. Although there are indications that the true corrosion rates are not simply proportional to these readings over the whole range, they demonstrate clearly that corrosion by H_2S-containing aqueous salt solutions is time-dependent. The time-averaged corrosion rates used in this paper cannot therefore be used to predict corrosion rates in practice.

The results of weight loss tests at 25 $^{\circ}$C in 0.1 and 10 % NaCl solutions with varying H_2S concentrations are given in Fig. 2. The weight losses observed in 0.1 % NaCl are apparently hardly dependent on H_2S concentration in the range 1 bar to

0.1 bar. At the low concentration of 10^{-3} bar the weight losses observed were very
small and it seems that lowering the H_2S pressure below ~ 0.1 bar decreases the
rate to an as yet undefined extent.

On the other hand, corrosion by H_2S in solutions containing 10 % NaCl
(corresponding to strongly saline formation water) was found to be dependent on
the H_2S pressure over a wide range. This difference in behaviour was reflected in the
type of iron sulphide layer formed. At low salt concentration this layer was compact,
almost shiny black in colour and difficult to remove. It may be similar to those
reported by Meyer et al. (Ref. 2) and Shannon and Boggs (Ref. 3), although the
above authors reported the formation of such layers only in the absence of salt. In
the tests with 10 % NaCl solution the layer of iron sulphide was porous and could
easily be removed. This agrees with the observations of Meyer et al. (Ref. 2) and
Shannon and Boggs (Ref. 3).

The very marked effect of the salt content of the medium is also evident
from the results of long-duration exposure tests given in Table I. The average
corrosion rate in 10 % NaCl (exp. d) was about an order of magnitude higher
than that in 0.01 % NaCl (exp. a) or 0.1 % NaCl (exp. b). At lower H_2S pressures the
salt effect seems less pronounced (Fig. 2), but proper comparison is difficult due to
the variation of H_2S solubility with salt concentration and a possible inhibitive
effect of chloride ions at higher concentrations.

The effects of H_2S pressure found in our work are more or less at variance
with those observed by Gardner (Ref. 7), Shannon and Boggs (Ref. 4) and Hausler
et al. (Ref. 6). Gardner (Ref. 7), working in a 4 % brine solution at 40 °C,
observed first-order dependence of corrosion rates upon H_2S pressure, and Shannon and
Boggs (Ref. 4), in 2 % salt solutions at 32 °C, found initial corrosion rates to
be dependent upon H_2S pressure. This dependence disappeared after 24 hours with
the build-up of a sulphide film. Finally, Hausler et al. (Ref. 6) found that in
10 % brine at 40 °C there was some dependence of corrosion rate upon H_2S concentrations from 0.02 to 0.25 bar.

As a pipeline may be fully rusted internally before its exposure to H_2S
in natural gas, a series of tests has been done in which the weight loss specimens
were pre-rusted in aerated salt solutions usually for four hours, before exposure
to H_2S. Although this pre-rusting gave some increase in the initial corrosion rates,
especially in the case of 0.1 % NaCl, the results in Table I show (exps. b,c,d,e)
that the average corrosion rates in long-term exposure experiments only increased
moderately. With pre-rusted samples the FeS layer formed in 0.1 % NaCl consisted
mainly of a dull black powdery layer which was readily removed, on top of a layer
tightly bound to the metal surface. Apparently, this layer still offered sufficient
protection. No effect of pre-rusting on the nature of the corrosion product
layers in 10 % NaCl was observed.

In a limited number of experiments flat corrosion coupons were rotated in
the test medium in such a way that the speed of the liquid along the sample
surfaces was nominally 1 m/s. The results for solutions containing 10 % NaCl are
also included in Fig. 2 and it is clear that under the conditions applied a speed
of 1 m/s has no significant effect upon the corrosion rates found. In a single
test with 1 bar H_2S in 0.1 % NaCl the average corrosion rate (0.29 mm/yr) was slightly
less than that found previously with slight stirring.

So far it would seem that at 25 °C with a high salt concentration neither
pre-rusting nor slight stirring markedly affect the corrosion rate, which maintains
a relatively high value. The corrosion product layer does not seem to give sufficient
protection and the corrosion found may be the maximum to be expected with
1 bar H_2S at this temperature in a nearly stagnant medium.

The influence of temperature on the average corrosion rates in 0.1 % NaCl
and 10 % NaCl is shown in Figs. 1 and 3. The results indicate that at high
temperatures a better protective layer is formed, presumably due to the reduced
solubility of iron sulphides at high temperatures (Ref. 8). This effect is most
pronounced in 10 % NaCl. It is also seen that increasing the stirring rate leads to

a faster build-up of the corrosion product layers. The more protective nature of these
layers in 10 % NaCl is also evident from the result of a long-duration experiment
(Table I, exp. f). X-ray diffraction has shown that the layers in this case consist
of mackinawite Fe_9S_8 and NaCl (see also Refs. 9, 10, 15).

THE CORROSION OF STEEL BY NaCl SOLUTIONS CONTAINING H_2S AND CO_2

The results of weight loss tests done in H_2S/CO_2 mixtures are shown in
Figs. 4, 5 and 6. The corrosion rate readings exhibited a time dependence similar
to those for corrosion by H_2S (see also Fig. 1). In 0.1 % NaCl at 25 °C the presence
of CO_2 in the gas mixture makes the corrosion product layer less protective and
easier to remove. It is seen that the average corrosion rate after 120 hours
exposure increases with increasing amount of CO_2 present (Fig. 4). Comparing the
average weight loss in 0.1 % NaCl at 25 °C in 0.2 bar H_2S, which is 31 mg after 308
hours, with that of 67 mg after 118 hours in 0.2 bar H_2S + 0.8 bar CO_2, clearly
demonstrates the role of CO_2 as corrosion promoter.

The effect of stirring speed on the formation of the protective layer,
which was discussed in the previous section, is also evident in these results. In the
tests in which the samples moved at a speed of 1 m/s through the 0.1 % NaCl solution,
lower corrosion rates were found (Fig. 4). In 10 % NaCl the situation seems to be
somewhat different (Fig. 5). The introduction of CO_2 in the gas mixture at 25 °C does
not affect the average corrosion rate to a great extent. When we compare Figs. 2 and
5, we see that the influence of CO_2 becomes only noticeable at lower p_{H_2S}/p_{CO_2} ratios.
Instead of decreasing with H_2S partial pressure, corrosion rates remain nearly constant over a wide range of p_{H_2S}/p_{CO_2} ratios and start rising at high CO_2 partial
pressures. It is clear that due to the unprotective nature of the corrosion product
layer, CO_2 can actively participate (presumably via the carbonic acid molecule
(Ref. 16)) in the corrosion of QMC41 steel in 10 % NaCl solution. As could be
expected from the discussion in the previous section, higher stirring rates did
not influence the corrosion rates at 25 °C under these circumstances (Fig. 5). The
results in 10 % NaCl at 80 °C agree with those reported for pure H_2S in the previous
section. A more protective sulphide layer is formed on the surface, thus reducing
the corrosion rates to somewhat lower values. This was also apparent from the
corrosion rate readings as a function of time; after initially displaying corrosion
rates well over 2 mm/yr, these readings dropped after 30-40 hours exposure to values
of the order of 0.2 mm/yr.

The results of some long-duration experiments at 25 °C in 0.1 % NaCl and
10 % NaCl are given in Fig. 6. In agreement with the results given above, the
corrosion rates in 0.1 % NaCl increase due to the presence of CO_2 in the corrosive
mixture. The corrosion product layers were found to be more protective and harder
to remove with increasing H_2S concentration. Apparently, upon introduction of CO_2
a mixture of iron sulphide and iron carbonate is deposited simultaneously on the
surface, thus reducing its coherence and protective qualities. General corrosion
rises from ~ 0.05 mm/yr in pure H_2S to ~ 0.2 mm/yr in H_2S/CO_2 mixtures containing
50 % and more CO_2. In 10 % NaCl the corrosion in 0.2 bar H_2S and 0.8 bar CO_2
(~ 0.3 mm/yr) was slightly lower than that in 0.5 bar H_2S, 0.5 bar CO_2 (~ 0.4 mm/yr).

The introduction of still more H_2S (0.8 bar) produced once again a
somewhat higher corrosion rate. This behaviour was accompanied by the formation of
increasingly more adherent corrosion layers and seemed to be at variance with the
results of the 120-hour exposure tests. However, after removal of the corrosion
products large spots of ring-like localized attack became visible in the case
of the long-duration experiments with the higher H_2S concentrations. It seems that
the more protective corrosion product layers, although offering some protection
against general attack, can also stimulate the appearance of much more dangerous forms
of localized corrosion.

It is clear that general corrosion in H_2S/CO_2 mixtures under conditions
such as can occur in transport and production lines will be dependent upon a complex
combination of at least the variables H_2S and CO_2 partial pressures, temperature,
flow rate of the aqueous phase and its salt content. More study is needed for a

better understanding of the phenomena which have been observed and before predictions
of the corrosion rates to be found in practice can be made from the variables quoted.

THE PITTING OF STEEL IN H_2S AND H_2S/CO_2 MIXTURES

As has been indicated in the introduction, pitting in H_2S and H_2S/CO_2-
containing solutions has received only limited attention (Refs. 11-14). It has
been claimed that sulphide layers develop noble rest potentials and can act as
cathodes in galvanic couples with steel resulting in accelerated pitting corrosion
of the steel. To study these effects, the variation of the rest potential of QMC41
steel with time under the influence of the build-up of a sulphide layer has been
followed in the extended tests reported in Table I and Fig. 6. Characteristic
results are given in Figs. 7 and 8. It can be seen that there is a pronounced
tendency for the iron sulphide covered steel in 0.1 % NaCl solutions to become
progressively more anodic with time. Anodic potential changes up to about 200 mV
were observed (Fig. 7, no. 1). No such shifts were observed at room temperature in
10 % NaCl, but at 80 ^{O}C anodic changes were found, although slightly less than in
0.1 % NaCl at room temperature. The gradual changes in electrochemical potentials can
presumably be explained by changes in the crystalline form and/or composition of the
iron sulphide, which are known to take place with time (Ref. 9).

The variations of the electrochemical potentials with time of samples in
0.1 and 10 % NaCl with different H_2S/CO_2 ratios are shown in Fig. 8. The rate at
which the samples achieve more anodic potentials with respect to bare steel is seen
to be dependent upon the quantity of CO_2 in the gas mixture. This is particularly
evident from the comparison of the curves 1 and 4, of which the latter displays
much higher potential values after much shorter exposure times. The presence of
CO_2 in the gas mixture pertaining to curve 1 clearly prevents the fast build-up of
a noble layer on the steel surface. If, as seems likely, the pitting of steel under
these conditions depends upon the development of a sufficiently large galvanic
couple between bare steel (the result of surface layer damage due to cracking,
erosion, etc.) and a protective corrosion product layer, then CO_2 has a positive
effect. It may reduce the rates at which highly anodic layers are formed and prevent
the occurrence of potentials exceeding values at which rapid film repair can no
longer take place. However, CO_2 can adversely affect the level of general corrosion,
as pointed out above. In the solutions containing 10 % NaCl at room temperature no
effect of CO_2 concentration was observed and no shifts of electrochemical potentials
with time occurred. Still, areas of local attack were observed at higher H_2S
concentrations, as discussed in the preceding section. More experiments will obviously
be needed to obtain a better understanding of the events taking place.

The simulation of pitting corrosion was attempted by short circuiting bare
steel electrodes (areas 0.5 and 0.05 cm^2) to a large steel plate (90 cm^2) covered
with a sulphide layer in an aqueous salt solution through which H_2S was bubbled. The
short circuiting was done via a zero-resistance ammeter, which was used to measure
the galvanic current flowing. In a second experiment a potentiostat was used to
maintain a zero-potential difference between sulphided iron reference electrode and
a bare steel measuring electrode.

A potentiostat was also used to immediately apply a chosen electrochemical
potential to a bare steel grit-blasted sample after its introduction into H_2S-
containing solutions. The potentials chosen were all anodic with respect to the
normal value for bare steel in these solutions; the anodic currents which flowed
were monitored against time. From these experiments it was clear that quite high
pitting rates can be possible. It was found that in a 1 to 100 anode to cathode area
ratio a penetration rate of 10 mm/yr (corresponding to a mean current density of
~ 8 µA/cm^2 on the cathode surface), only causes a potential shift of ~ 7 mV for the
iron sulphide layer.

In the majority of the pitting simulation tests done so far, the "pits"
lost their activity within a few days as a result of the build-up of protective
corrosion product layers on the bare steel surface. Since in practice pits retain
their activity long enough to penetrate pipe walls of considerable thickness more

experiments, better suited to represent the complex steady state of a pit in practice, are required. Investigations of this type are now being considered.

CONCLUSIONS

The influence of salt content of a solution in which corrosion of steel by H_2S or a H_2S/CO_2 mixture is taking place has clearly been demonstrated. At 25 °C high salt concentrations prevent the formation of protective surface layers that could stifle further attack, and relatively high general corrosion rates are observed. At higher temperatures (80 °C) better protective layers are formed, even in the case of high salt concentrations.

Introduction of CO_2 in the H_2S corrosion in 0.1 % NaCl solutions has been shown to increase the general corrosion rates.

The salt content of solutions also influences the development of protective surface layers of corrosion products on steel, the electrochemical potential of such layers and hence their possible galvanic activity. The development of highly anodic surface layer potentials is dependent upon temperature and also upon the H_2S/CO_2 ratio.

ACKNOWLEDGEMENT

We thank Ms. J. van den Tol for her valuable assistance with the experimental work.

REFERENCES

1. Proc. 2nd International Congress: Hydrogen in Metals. Organised by the International Association for Hydrogen Energy. Paris, June 6th-10th, 1977, Pergamon Press.

2. Meijer, F.H., Riggs, O.L. McGlasson, R.L., and Sudbury, J.D.: "Corrosion Products of Mild Steel in Hydrogen Sulfide Environments". Corrosion, 14, pp. 109t-115t. (1958).

3. Shannon, D.W. and Boggs, J.E.: "Factors affecting the corrosion of steel by oil-brine-hydrogen-sulfide mixtures". Corrosion, 15, pp. 299t-302t. (1959).

4. Shannon, D.W. and Boggs, J.E.: "An analytical procedure for testing the effectiveness of hydrogen sulfide corrosion inhibitors". Corrosion, 15, pp. 303t-306t. (1959).

5. Newman, T.R.: "A laboratory method for evaluating corrosion inhibitors for secondary recovery". Corrosion, 15, pp. 307t-310t. (1959).

6. Hausler, R.M., Goeller, L.A., Zimmerman, R.P. and Rosenwald, R.H.: "Contribution to the "filming amine" theory: an interpretation of experimental results". Corrosion, 28, pp. 7-16. (1972).

7. Gardner, G.S.: "Reaction velocity in the corrosion of iron by hydrogen sulfide and the effect of inhibitors". Corrosion, 16, pp. 312t-318t. (1960).

8. Pohl, H.A.: "Solubility of iron sulfide". J. Chem. Eng. Data, 7, pp. 295-306. (1962).

9. Smith, J.S. and Miller, J.D.A.: "Nature of sulfides and their corrosive effect on ferrous metals. A review". Brit. Corrosion J., 10, pp. 136-143. (1975).

10. King, R.A. and Miller, J.D.A.: "Corrosion of ferrous metals by bacterially
produced iron sulfides and its control by cathodic protection". Proc. First
International Conference on the Internal and External Protection of Pipes.
Paper F2. Organised by BHRA Fluid Engineering. University of Durham
(September 9th-11th, 1975).

11. Klas, H.: "Contribution to the solution of corrosion problems occurring in oil
field piping carrying iron sulfide bearing brine". Sixth World Petroleum
Congress. Paper 8, Section VII. pp. 19-26. Frankfort/Main (June 19th-26th, 1963).

12. Naumann, F.K. and Carius, W. : "The importance of the corrosion processes in
aqueous hydrogen sulfide solutions". Arch. Eisenhüttenw., $\underline{30}$,
pp. 283-292. (1959). (in German).

13. Sandoz, G., Fujii, C.T. and Brown, B.F.: "Solution chemistry within stress -
corrosion cracks in alloy steels". Corrosion Science, $\underline{10}$, pp. 839-845. (1970).

14. Akol'zin, P.A. and Bogachev, A.F.: "Corrosion of carbon steels in water
containing hydrogen sulfide and sulfur". Teploenergetika, $\underline{15}$, pp. 33-36.
(1968). (in Russian).

15. Yamaguchi, S. and Moore, T.: "On the formation of cubic Fe_3S_4 during corrosion
of steel". (Zur Bildung der Kubischen Fe_3S_4 bei der Korrosion von Stahl).
Werkstoffe und Korrosion, $\underline{24}$, pp. 28, 29. (1973). (in German).

16. De Waard, C. and Milliams, D.E.: "Prediction of carbonic acid corrosion in
natural gas pipelines". Proc. First International Conference on the Internal
and External Protection of Pipes. Paper F1. Organised by BHRA Fluid Engineering.
University of Durham (September 9th-11th, 1975).

Table I: Average corrosion rates of QMC41 steel in 1 bar H_2S

Exp.	Electrolyte (% NaCl)	Condition	Temperature ($^\circ$C)	Exposure time (h)	Corrosion rate (mm/yr)
a	0.01	grit blasted	25	5087	0.07
b	0.1	grit blasted	25	4363	0.04
c	0.1	pre-rusted	25	4363	0.06
d	10	grit blasted	25	4363	0.33
e	10	pre-rusted	25	4363	0.58
f	10	grit blasted	80	1920	0.11

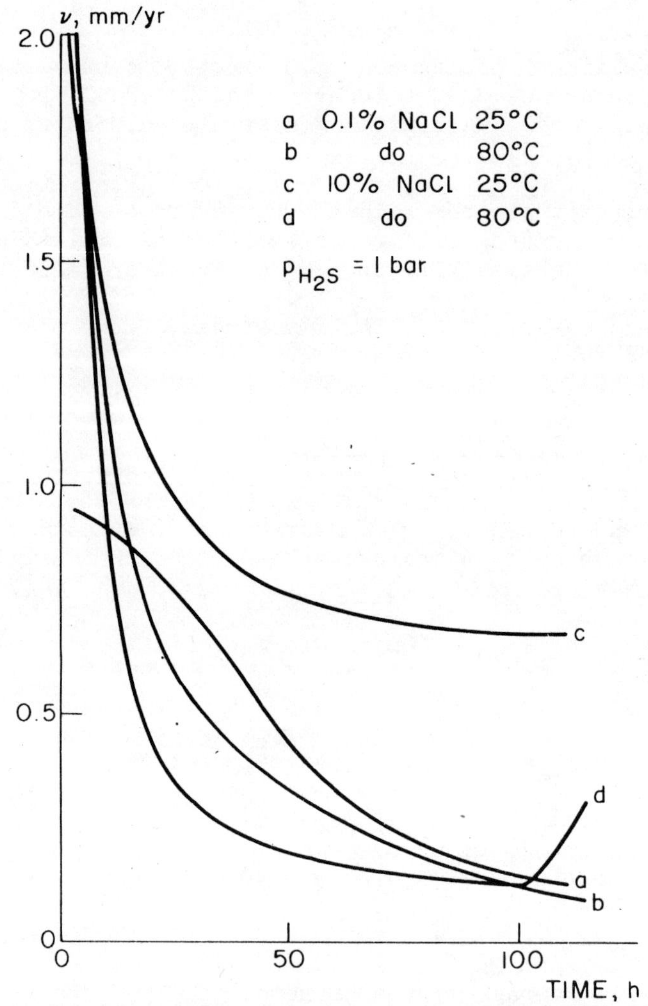

Fig. 1   Corrosion rate readings for QMC41 steel against time.

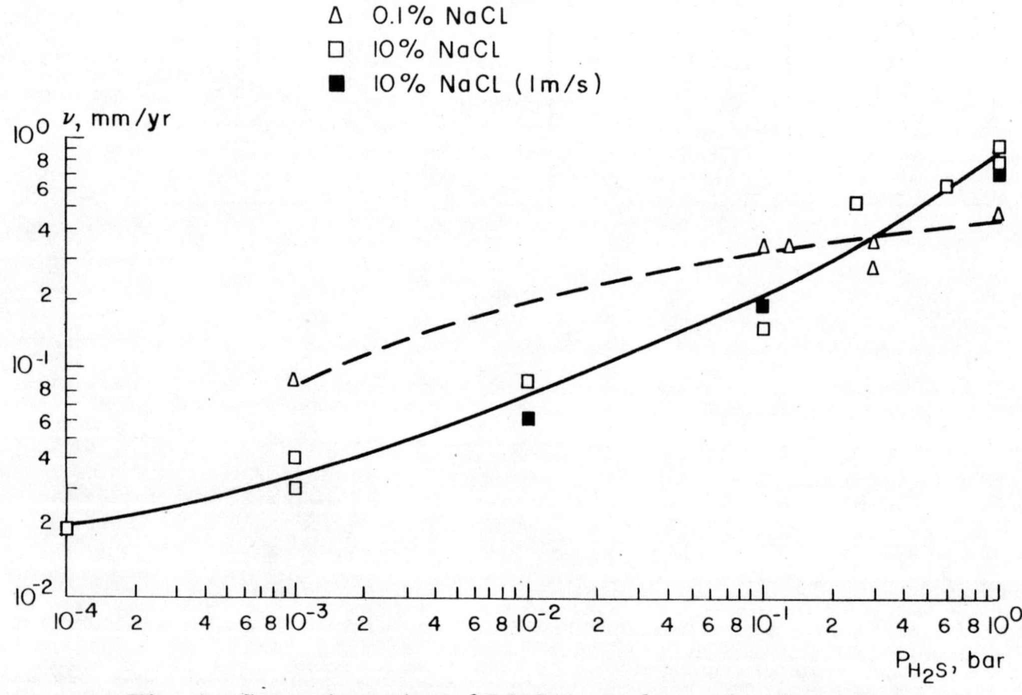

Fig. 2   Corrosion rates of QMC41 steel as a function of H_2S
partial pressures for various NaCl solutions (T = 25 °C).

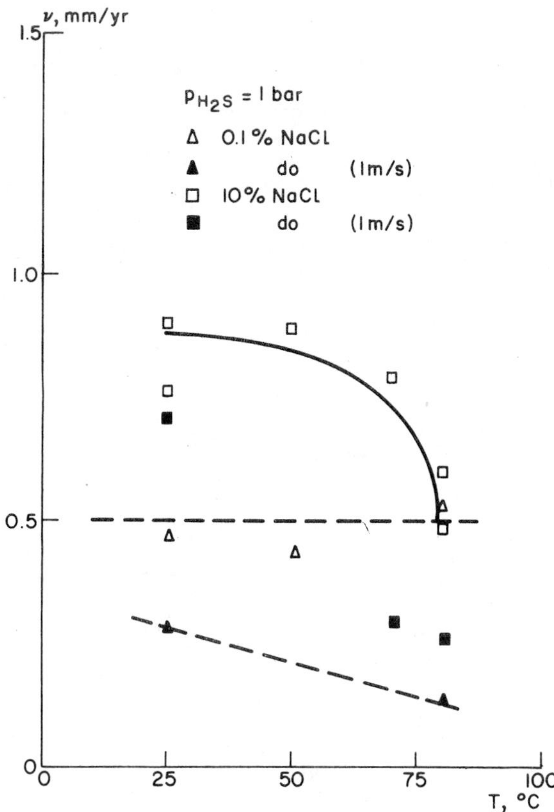

Fig. 3   Corrosion rates of QMC41 steel as a function of temperature.

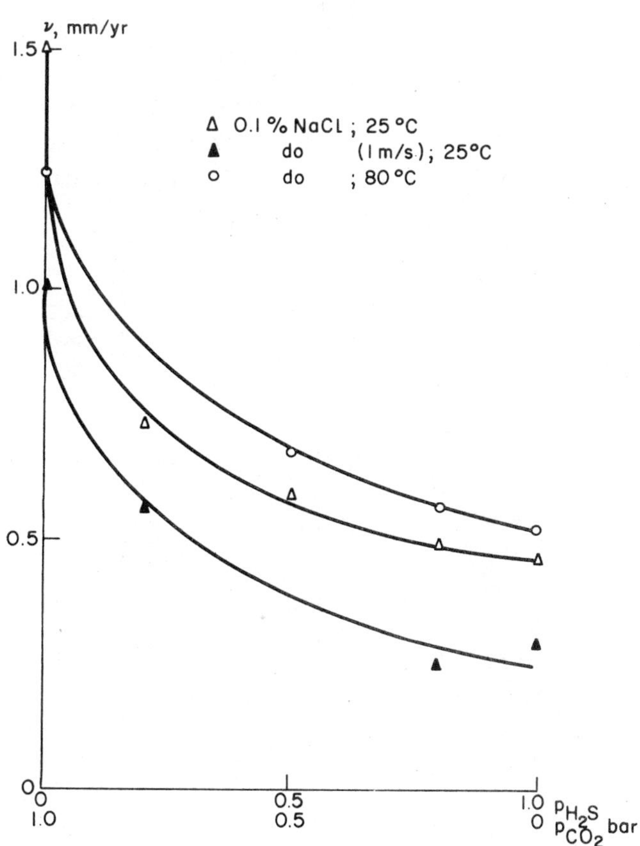

Fig. 4   Corrosion rates of QMC41 steel as a function of H₂S and CO₂ partial pressures.

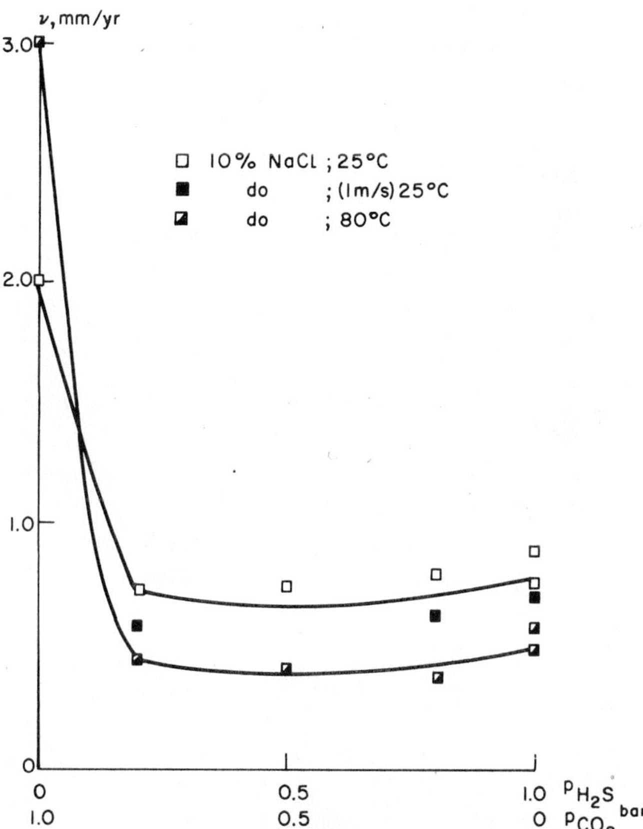

Fig. 5   Corrosion rates of QMC41 steel as a function of H₂S and CO₂ partial pressures.

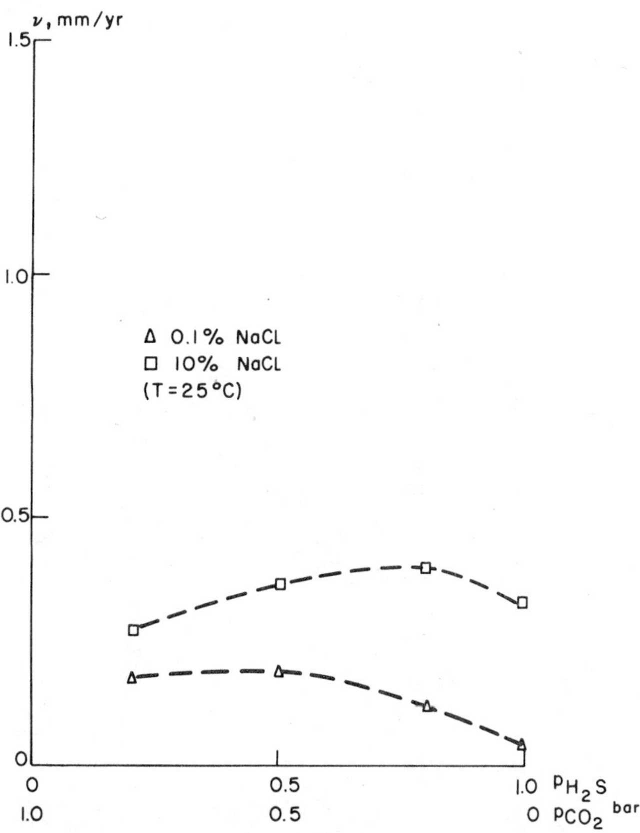

Fig. 6   Corrosion rates of QMC41 steel after 4824 hours exposure as a function of H₂S and CO₂ partial pressures.

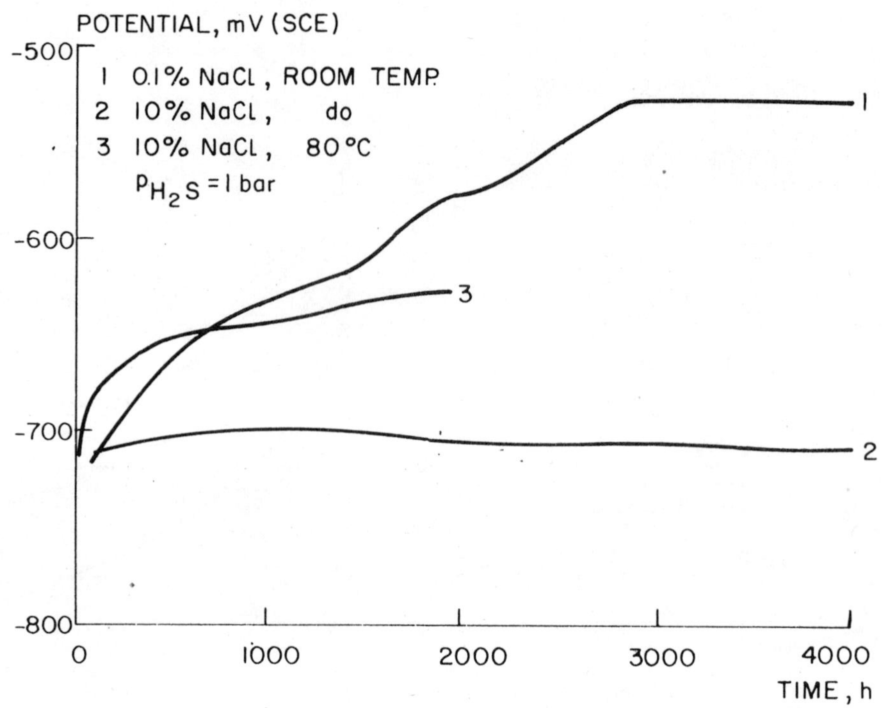

Fig. 7   Electrode potentials of QMC41 steel as a function of time.

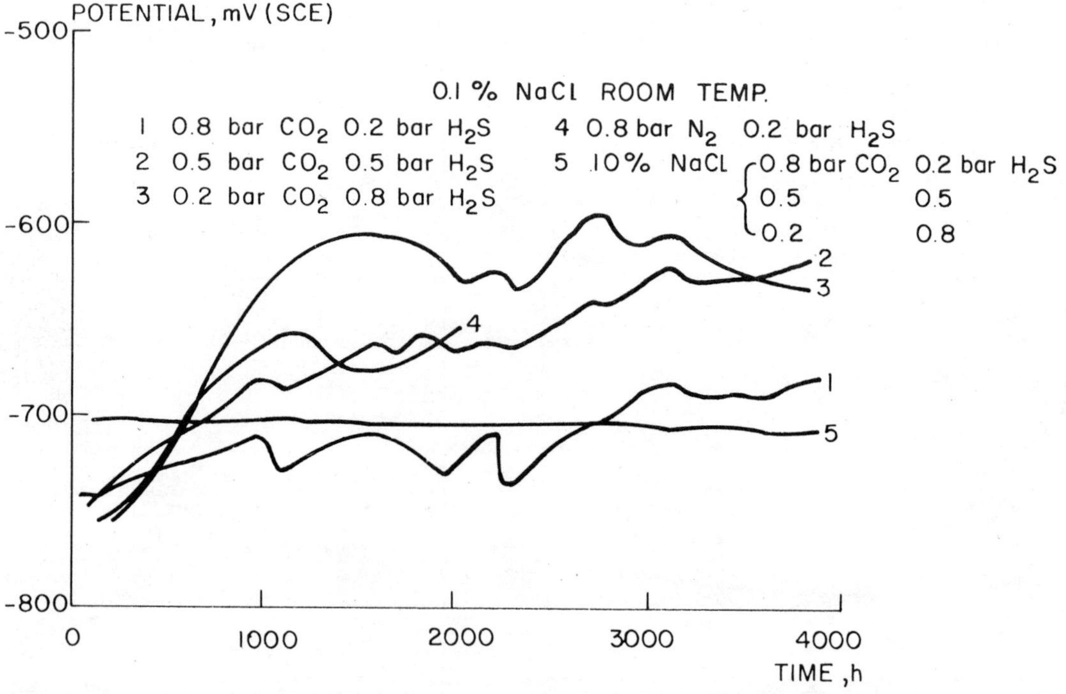

Fig. 8   Electrode potentials of QMC41 steel as a function of time.

internal and external
protection of pipes

September 5th - 7th, 1979

A NEW AND IMPROVED JOINT COVER SYSTEM FOR PIPELINE

K. Yamaguchi, S. Ogiwara, T. Nagasawa and I. Tsurutani.

Ube Industries Ltd., Japan.

Summary

This paper relates to joint covers which protect the welded part of the pipeline from external corrosion. Firstly, the constitution of two types of joint cover, one an improved torch type joint cover sheet installed by torch flame and the other a newly developed heater-included type joint cover installed by application of electric power are described. Secondly, various problems of these two types of joint cover at installation are discussed based on the experimental data. On the selection and installation of the simple and low cost torch type joint cover, the following problems must be carefully considered.

The adhesive strength of the joint cover to the pipeline depends on the pipe temperature at the installation. However the long time heating by torch flame to raise the pipe temperature causes degradation or oxidation of the joint cover surface, which induces the mechanical properties to make the joint cover worse. The heater-included joint cover has many advantages including shorter installation time under severe weather, uniform and fine finishing, strong adhesive strength.

For the durability of the pipeline, the combination of raw material polyethylenes of mill coat and joint cover must be carefully considered.

Held at Imperial College, London, England.
Organised and Sponsored by BHRA Fluid Engineering
Copyright BHRA Fluid Engineering, Cranfield, Bedford, England.

1. INTRODUCTION

The following pipe coating materials to protect the pipe from external corrosion are commonly used: bituminous material, epoxy resin, and polyethylene resin. Though the bituminous material has been used the longest of the three, recently, use has been made of other materials, because of the temperature dependency of the bituminous material, viscosity brought outflow at high temperature or brittleness at low temperature. Epoxy resin is mainly used in the USA, and this has excellent hardness. But the thickness of the coating layer is very thin, thus it is easily cracked by impact or bombardment and bending of pipes at installation or handling as well as generating many pinholes during the coating process, which are very important in the long term reliability and life of the pipeline. As the polyethylene coating layer is comparatively thick, about several millimeters, there is no chance of generating pinholes and its flexibility together with thickness make it difficult to be cracked by impact or bending during transportation or installation. Also its excellent insulating properties are useful for electro chemical anticorrosion of the pipeline.

Two polyethylene coating methods have been developed: Polyethylene tape, with sticky adhesive wrapped around the pipeline at the installation site, and molten polyethylene sheet with adhesive coated around the pipes in the mill (mill coated pipe). It is considered that the latter case is favourable due to the adhesive strength of polyethylene to the pipe surface and its stable anticorrosive property. As the mill coated pipes, which are usually ten to fifteen metres long, are interconnected at the installation site by welding, about several tens of centimetres of the connecting part must be coated by anticorrosive protection at the site after welding. Materials for coating the welding part are defined as a joint cover in this paper. The schematic sectional diagram of the welding part and installed joint cover are illustrated in Figure 1. Mechanical and physical properties of the installed joint cover must be equal or superior to those of the mill coat layer, or else the excellent anticorrosive performances and durability of the mill coat layer come to naught in the total pipeline system. Consequently very severe performances are required of the joint cover together with reliable and rapid installation time as follows:

(1) Installation of joint covers

    (1-1) The installation time must be within several minutes, even for a large diameter pipeline.

    (1-2) A tube type joint cover which must be set on the pipeline before welding and shifted to the welding part is not desirable, but a sheet type joint cover which can be set after welding is desirable.

    (1-3) The installation of the joint cover can be done under any weather conditions, at very low temperatures even below $-30^{\circ}C$ or in rain, snow and wind.

    (1-4) It is desirable that performances of the installed joint cover do not depend on the degree of proficiency of operators.

(2) Performances of the installed joint cover.

    (2-1) Stable and sufficient adhesive strengths of the joint cover to the pipe surface near the weld line and the joint cover to the mill coated polyethylene at a temperature range of $80^{\circ}C$ to $45^{\circ}C$ are required.

    (2-2) The durability and the anticorrosive effects of the installed joint cover must be equal or superior to those of the mill coated polyethylene.

    (2-3) Water or harmful gas must be sealed for a long time even under high static pressures as at the deep sea bottom.

Considering the above severe requirements, the method of installation and performances of the installed joint cover of the improved torch type joint cover and the newly developed heater included type joint cover will be explained and discussed in this paper.

2. DESCRIPTION OF TWO TYPE JOINT COVERS

(1)   Torch type joint cover

The composition of a torch type joint cover used in this paper and the setting manner of the cover to a pipe for installation is shown in Figure 2. The cover consists heat shrinkable polyethylene sheets and adhesive material and both ends of the sheet are separated into two layers.

In experiments the thickness of the polyethylene layer was varied from 0.9 mm to 3.2 mm and the thickness of the adhesive material fixed at 1.0 mm. Both separating ends of the joint cover were engaged, as shown in Figure 2, at the top of the pipe and a sticky seal tape was put on the joint cover end, which was taken off after installation.

(2)   Heater-included type joint cover

This type of joint cover consists thermal shrinkable polyethylene sheet in which electric heaters are included with adhesive materials. Voltage applied to electric terminals in the cover causes current flow through electric heaters in the joint cover, the temperature of the cover then rises gradually and the cover starts to shrink. A schematic diagram of the composition of the heater included type cover is shown in Figure 3. Electric heaters are, arranged at almost right angles to the shrink direction with folds as shown in Figure 3. Calorific value and electric current are controlled by applied voltage and resistance of heaters according to the following equations. When the electric resistance of a unit length (1 cm) heater is r (ohm), the total length of a heater lo (cm) is expressed by the width W (cm), length L (cm) of the joint cover, pitch of adjacent heaters D' (cm), and number of heaters N in the joint cover.

$$lo = \frac{L \cdot W}{D' \cdot N} \quad (cm) \tag{1}$$

Then, the resistance of whole heater, R (ohms), is

$$R = \frac{L \cdot W}{D' \cdot N} \, r \quad (ohms) \tag{2}$$

The resistance R increases with the temperature rise of the heater. When the applied voltage is V (volt), the initial total electric current I_s and watt density per unit (1 cm^2) area of the cover w_s (watt/cm^2) are,

$$I_s = \frac{V \cdot N}{R} = \frac{V \cdot D \cdot N^2}{L \cdot W \cdot r} \tag{3}$$

$$w_s = \frac{V \cdot I}{L \cdot W} = \left(\frac{V \cdot N}{L \quad W}\right)^2 \cdot \frac{D}{r} \tag{4}$$

Using these relations, the number of heaters (N), pitch of adjacent heaters (D'), and the resistance of heater (r) for the demanded size of width and length of a joint cover are determined and the joint cover manufactured.

3. INSTALLATION

(1)   Test for installation

Pipes used in experiments for installation are as follows (the corresponding symbols are shown in Figure 1):

Diameter of pipe (Pd, mm)	250
Thickness of pipe (St, mm)	7
Width of welding part (W$_w$, mm)	270
Mill coat polyethylene thickness (Ct, mm)	2.5

A fine thermistor to measure the temperature rise of the pipe during installation was set up at the lowest point of the welding line, Pt in Figure 4, and another thermistor to measure the temperature rise of the joint cover was attached on the side surface of the joint cover, Jt in Figure 4. Size and specification of joint covers used in the following experiments are shown in Table 1 (symbols corresponding to these are shown in Figures 2 and 3). Specifications for heaters in the heater included joint cover are listed in Table 2.

Table 1. Size of joint cover

Item	Torch type cover	Heater included type cover
Width (Jw, mm)	450	450
Length (L, mm)	1,000	1,000
Separating end length (DL, mm)	80	80
Thickness of adhesive (d_2, mm)	1	1
Thickness of polyethylene sheet (d_1, mm)	0.95 – 2.85	1.25 – 3.20

Table 2. Specification of heaters

Number of heaters (N)	1*
Resistance of a unit length heater (ohms/cm)	3.1×10^{-3}
Pitch of adjacent heaters (D', mm)	3 – 8
Applied Voltage (V, Volts)	80 – 200
Initial Watt density w_s (watt/cm^2)	0.25 – 1.0
Temperature at installation (^{O}C)	-22^OC – 37^OC

* For practical use 5 and 8 are used to make one electric terminal.

Electric heaters are always embedded in the centre of polyethylene sheet section. To check the finish time of installation, a protrusion was made on the bottom surface of a pipe, Ce in Figure 4, and the point which needed to be clearly observed was the shape of the protrusion on the shrunk joint cover at the installation and thus was defined as the finishing time of the installation.

(2)   Test for torch type joint cover
     A torch type joint cover was set on a pipe as shown in Figure 2 and was heated by a torch of propane gas from top to bottom of a centre welding line and both sides in turn.

Test results of three different thicknesses of joint cover consisting of various thicknesses of polyethylene sheet indicated by t in Figure 4 and 1 mm thick adhesive material installed by torch flame are shown in Figure 4.

The initial pipe temperature was 40°C and atmospheric temperature was 25°C. The joint cover was heated from top to bottom. Heat transferred to the top of the joint cover to the top of pipe. The pipe bottom temperature started to rise after two minutes, thus time hardly depended on the thickness of the pipe. The increase of the pipe temperature saturated around 60°C with any thick joint cover. All kinds of cover were perfectly shrunk and beautifully completed within six to ten minutes, but the pipe temperature did not increase any more by longer heating after it had finished shrinking. This result means that the transfer of heat from the joint cover to the pipe is equal to the transfer of heat in the pipe along the pipe line direction. This saturated temperature depends on the pipe thickness and atmospheric temperature, but hardly depends on the thickness of the joint cover and strength of the torch flame or heating time. This pipe temperature at installation is very important to the adhesive strength of the joint cover to the pipe surface as described later.

(3)   Test for heater-included joint cover

The setting for a heater-included joint cover to the pipe before installation is the same as for a torch type joint cover except for the connection of the power supply.

A settled voltage in advance adjusted by a transformer was applied to the heater and the watt density of the cover decreased with time due to the increase of the resistance of the heater with higher temperature. Using this phenomena the temperatures of heaters were estimated to check the degree of thermal degradation or oxidation of the polyethylene sheet.

One typical result of the installation of this type of joint cover was where the initial pipe temperature was 45°C and the initial watt density was 0.75 watt/cm^2 and this is shown in Figure 5. The pipe temperature started to increase after two minutes and rose to around 90°C. The shrinkage of the joint cover was complete about four minutes after; the increase of the pipe temperature continued with application of electric power.

The temperature of the whole joint cover increased all at once.

When the joint cover contacts the pipe surface by shrinkage, heat transfer to the pipe and the cover temperature decreases temporarily and again increases. From the results in Figure 5 it is shown that installation was completed with 3.5 minutes application of electric power, where the joint cover temperature was 90°C and the pipe temperature was 46°C. When the electric power was continuously applied for 9 minutes, the cover temperature reached 103°C and the pipe temperature reached 80°C. The pipe temperature can thus be controlled and is very useful for the adhesion strength of the pipe to the joint cover.

The pipe temperature increases drastically after completion of the shrinkage as shown in Figure 6.

Firstly the installation times, that is, the times for completion of the shrinkage were checked with various thick joint covers and power applied conditions.

Installation times of both torch type and heater included type joint covers with various thicknesses were measured as shown in Figure 7. Installation time increases linearly with thickness of both types of joint cover and the installation time of the heater-included type covers is almost half that of torch type covers with the same thickness. The results in Figure 7 were obtained by the 250 mm diameter pipe, the difference and the ratio of installation times of those two types of cover increase with the larger diameter pipe. That is, the merits of the heater-included type joint cover increase with larger diameter pipeline. Installation times of the heater included type joint cover depend on several other conditions - watt density, pitch of adjacent heater, and atmospheric temperature as shown in Figure 8, which suggests that 3 to 5 mm pitch of adjacent heaters is adequate and the watt density must be selected by atmospheric temperature and the desirable installation time. A second problem is the possibility of thermal degradation or oxidation of polyethylene sheet by heaters.

Applying a constant voltage to heaters, the real watt density of the joint cover decreases with time as shown in Figure 9, caused by the increase of heater resistance with higher temperature. From these data the temperature of the heater can be calculated as shown in Figure 10. When the initial watt density is 1.0 watt/cm^2, the real watt density decreases to 0.7 watt/cm^2

2 minutes after, the cover completes the shrinkage and the heater temperature is 120°C.
When the cover is continuously heated for 6 minutes with the same voltage, the real watt density becomes 0.6 watt/cm^2 and the heater temperature is 160°C. As the thermal degradation[1] or oxidation[2] occurs over 200°C in the range of installation time, there is no need to take this into consideration for the heater-included type joint cover at installation.

The third problem is the atmospheric temperature at installation. Installation times of two types of joint covers under various atmospheric temperatures were measured as shown in Figure 11.

On the heater-included type joint cover 2.2 minutes installation time at 36°C increases a little to 3.6 minutes at -2°C with the same applied voltage, otherwise on the torch type joint cover, 8 minutes installation time at 36°C increases to 13 minutes. This installation time of the heater-included type joint cover is hardly effected by weather conditions, rain, snow, or wind; on the other hand, the torch type is strongly effected by those conditions. This means that the heater-included type joint cover is very useful under severe atmospheric conditions.

(4)   Computer simulation of the installation process.

The heat generation and transfer process inside the heater-included type joint cover at the installation were analysed by computer simulation to establish the installation standard manuals. A co-ordinate of a cross-section of a joint cover for the numerical calculation is shown in Figure 12. Heat was generated from heat surface transfer in the joint cover and heat release in the air from the surface. The temperature at A in Figure 12 is defined as the joint cover temperature. Calculated values approximately agree with experimental values within three minutes as shown in Figure 13. In the practical installation the joint cover contacts tightly to the pipe for around three minutes caused by shrinkage of the cover and heat transfer to the pipe to make the cover temperature fall. On the other hand the calculated joint cover temperature never falls during the installation because the joint cover never comes in touch with the pipe surface on the calculation. The temperature distribution in the cover during the installation is shown in Figure 14. The temperature around the heater is a little higher than that in the other part, but the difference at the heater and at the lowest part is minimal.

4. MECHANICAL AND PHYSICAL PROPERTIES

Similar or superior mechanical and physical properties are required for the installed joint cover than for those of mill coat polyethylene to ensure excellent anticorrosive performance of the total pipeline. However these requirements for the installed joint cover are very severe because of the many restricted conditions at installation.
Some important properties required of the installed joint cover are as follows.
(A)   The adhesive strength of the installed joint cover to the pipe surface is important due to cathodic disbonding for anticorrosive protection and the strong shear stress from the tensioner maintaining the pipe at the installation.
(B)   Adhesive strengths of the installed joint cover to the mill coat polyethylene and of the contact of both joint cover ends to prevent the permeation of sea water or rain water are important.
(C)   Mechanical properties and durability of the installed joint cover, especially environmental stress cracking resistance, tensile break strength, elongation, yield strength, and impact strength at low temperature, must be equal or superior to those of the mill coat polyethylene.
The properties of two installed joint covers were analysed.

(1)   Adhesive strength
   (1-1)   Adhesive strength to steel pipe surface

Two kinds of adhesive, hard and soft, are used to adhere the mill coat polyethylene to a steel pipe surface. Soft adhesive consists mainly of bituminous materials or synthetic rubber materials and their sticky properties are used for adhesion. These adhesives have the advantage of being charged in every gap or pinhole and have an excellent hermetic property, but their fluidity is very sensitive to temperature which decreases the adhesive strength and flows out at high temperatures. Some of these materials become very brittle at low temperatures, which effect the polyethylene coating layer body. On the other hand, hard adhesive consisting of chemical modified polyethylene or polyamide resin, which make chemical bonds to the steel, have almost equal properties to ordinary polyethylene. For a stable adhesive strength in a wide temperature range and for an excellent cathodic disbonding property, it is felt that the hard adhesive is better for attaching mill coat polyethylene to steel pipe. One problem of the hard polyethylene is the high temperature of the adhesive and steel pipe during adhering. On the mill coat process the pipe should be kept at the right temperature by preheating.

As previously mentioned, the rise of pipe temperature at the installation of a torch type joint cover is saturated at about 60°C; therefore it is very difficult to use the hard adhesive for the torch type joint cover. Thus, an intermediate type of adhesive was developed for the torch type joint cover. Adhesion strength at room temperature of the adhesive depends on the temperature while adhering as shown in Figure 15. Therefore the pipe temperature at the joint cover installation must be determined by the demanded adhesion strength. If a strong adhesion strength is demanded to make the shear deformation withstand the tensioner stress during the installation as in a deep sea pipeline, the heater type joint cover which has stronger adhesion properties is favourable.

(1-2)   Adhesive strength to mill coat pipeline

The strong adhesion between the joint cover and the mill coat polyethylene for strong shear stress caused by the thicker part, as shown in Figure 1, and the pipe is hermetically sealed from sea water or rain water. The melting temperature of mill coat polyethylene is in the range of 95 to 110°C for high pressure processed polyethylene and 110 to 125°C for middle or low pressure processed polyethylene. Oriented and irradiated polyethylene sheet is used for the joint cover of which the melting temperature is about 90 to 110°C.

Originally polyethylene material was very hard for use in molten adhesion because of its own tremendous high viscosity at the molten state. Therefore the mill coat polyethylene with a lower melting temperature and the joint cover cosisting the same kind of polyethylene to the mill coat polyethylene is favourable for sealing and adhesion. If the same kind of polyethylenes are used for mill coat and joint covers, it is very favourable for mechanical resistance during bending or cyclical temperature fluctuation for drastically improving the total anticorrosive properties.

Namely, in the selection of joint covers, a combination of all sorts of polyethylene mill coat and joint covers must be considered as well as the installation method.

(1-3)   Adhesive strength of both ends of a joint cover

Though a tube type joint cover does not need a connection, it must be set before the welding of pipes and moved to the welding part. If the installed joint cover is not perfect caused by either misses during installation or original defects of the joint cover, it is very difficult to repeat the installation. Therefore a sheet type joint cover which can be set after welding is favourable and many connecting methods for sheet ends are considered and used. One of these methods is shown in Figure 2, where the thicknesses of individual sheets are not equal and the stronger adhesion strength is determined by considering the thermal conductivity, melting temperature, and viscosity of molten state polyethylene.

Long time sealing tests and cathodic disbonding tests for installed pipes by two type joint covers under hydro-static pressure of sea water corresponding to a 700 metre deep sea are continued and the results will be discussed at the conference.

(2)   Mechanical properties

(2-1)   Environmental stress cracking resistance

Firstly this property was discussed on the polyethylene sheet material of power or comunication cables for long term durability. Cracks are generated on the inner stressed polyethylene material in air, water, or underground and the test method promoted to check this property of polyethylene is regulated by ASTM D1993.

The time after 50% failure of specimens on the test defined as ESCR, F50 depends on the melting flow index (MI) of polyethylene and the manufacturing process as shown in Figure 16. These results mean that the original polyethylene material of the joint cover which has good ESCR must be carefully considered and selected. The surface of the torch type joint covers are

burned by intense torch flame at installation. The thermal degradation or oxidation of
polyethylene must have affected the ESCR badly. The degradation or oxidation to the whole
polyethylene sheet is negligible, but it is generated only on the surface and is detected by the
surface infra-red measurements as shown in Figure 17. As burning times of specimens in
Figure 17 are almost twice those of the ordinary installation, the ESCR of polyethylene becomes
worse. This result suggests the effect to ESCR must be considered at the installation of the
torch type joint cover. On the other hand the change in surface degradation of the heater type
joint cover at installation cannot be detected by infra-red measurement and the ESCR is not
made any worse.

(2-2)   Mechanical properties

The joint cover consists of crosslinked shrinkable polyethylene and the degree of
shrinkage of a torch type joint cover varies locally top or bottom of the pipe or, at the centre
of the sides of the joint cover. This irregular shrinkage causes unevenness of thickness and
inequality of mechanical properties.
Effects of the low temperature brittleness of the residual degree of shrinkage, which was defined
as shown in Figure 18, was checked as shown in Figure 19. Though the low temperature of
brittleness for every un-notched specimen is below $-60^{\circ}C$, 0.2 mm depth notched specimens
depend on the residual degree of shrinkage and direction of the joint cover.

Elongational amounts and strengths at breaking and yield stresses along the
shrinking direction and its perpendicular direction on the tensile elongation tests of various
residual degree of shrunken specimens at $-45^{\circ}C$ and $20^{\circ}C$ are illustrated in Table 3.

These uneven residual shrinkage effects are not so sensitive to mechanical
properties; however it is important that there is uniform shrinkage at installation.

Table 3

Degree of Residual Shrinkage		40%			20%			0%		
Direction		Ra*	Ax**	Ratio	Ra	Ax	Ratio	Ra	Ax	Ratio
Yield Strength (kg/cm^2)	at $-45^{\circ}C$	320	380	1.19	340	370	1.29	350	360	1.03
	at $20^{\circ}C$	90	120	1.33	90	110	1.22	100	100	1.00
Break Strength (kg/cm^2)	at $-45^{\circ}C$	480	400	0.83	460	400	0.87	400	400	1.00
	at $20^{\circ}C$	360	300	0.83	310	290	0.94	230	240	1.04
Elongation (%)	at $-45^{\circ}C$	360	470	1.31	380	430	1.13	420	420	1.00
	at $20^{\circ}C$	440	510	1.16	460	520	1.13	510	500	0.98

*   Ra : Circumference direction to the pipe

** Ax : Axial direction to the pipe

5. CONCLUSIONS

Some problems of joint covers used for the external corrosion protection of the joint welding part of a pipeline were discussed from the view point of mechanical and physical properties of installed joint covers and at installation. Durability and protection from corrosion performance of installed joint covers must be equal or superior to those of mill coat polyethylene. Mechanical and anticorrosive characteristics of two types of installed joint cover, their desirable installation methods, and standards to select a joint cover are summarized from the above results.

(A) Constitution

Sheet type joint covers which can be installed after the welding of the pipeline, are more desirable from the installation angle than tube type joint covers which must be set before welding. The installed joint cover must prevent the permeation of sea water or rain water from the contact part of both ends of the sheet or the interface of the joint cover and the mill coat polyethylene. Prevention of water permeation of the installed joint cover under high hydrostatic pressure must be experimentally checked.

(B) Installation

Longer heating time by flame during the installation of the torch type joint cover is not desirable so as to cause the thermal degradation or oxidation of the joint cover surface.

At installation the degree of shrinkage in a torch type joint cover must be uniform to avoid the fluctuation of thickness and make the mechanical properties worse.

The heater-included type joint cover has the following advantages.

(1) Degradation or oxidation do not occur on the joint cover surface.

(2) There are no fluctuations in thickness and degree of shrinkage.

(3) It can be installed in any weather; very low temperature, rain, snow or wind.

(4) Mechanical and physical properties of the installed joint cover do not depend on the technical knowledge of the operator.

(5) Installation time, especially for large diameter pipelines, is very much shorter than that of the torch type joint cover.

(6) Pipe temperatures at installation which strongly effect the adhesion strength of the joint cover to the pipe can be controlled at one's will.

(7) As installation time dependency on the thickness of the joint cover is lower than that of the torch type joint cover, a thicker joint cover can be used.

(C) Mechanical and physical properties

If the quality assurance to protect the installed joint cover from external corrosion is not sufficient, the high quality assurance of the mill coat polyethylene comes to naught. Therefore the environmental stress cracking resistance (ESCR), low temperature brittleness, or mechanical properties of the installed joint cover must be carefully considered. The combination of low material polyethylenes of mill coat and the joint cover to prevent the permeation of sea or rain water must be considered.

The remaining heater of the heater included type joint cover after installation has no real effect on the anticorrosive protection. The thickness of the heater-included type joint cover is thicker and some types can be peeled off the heater after installation. The protection quality of the remaining installed cover without heaters is perfect and it equals those before the peeling. This result means that the leaving of the heater in the joint cover does not have any effects on the protection ability of the joint cover.

REFERENCES

1) L. A. Wall, S. Straus ; J. Polym. Sci., $\underline{44}$, 315 (1960)

2) K. B. Chakraborty, G. Scott ; European Polym. J., $\underline{13}$, 732 (1977)

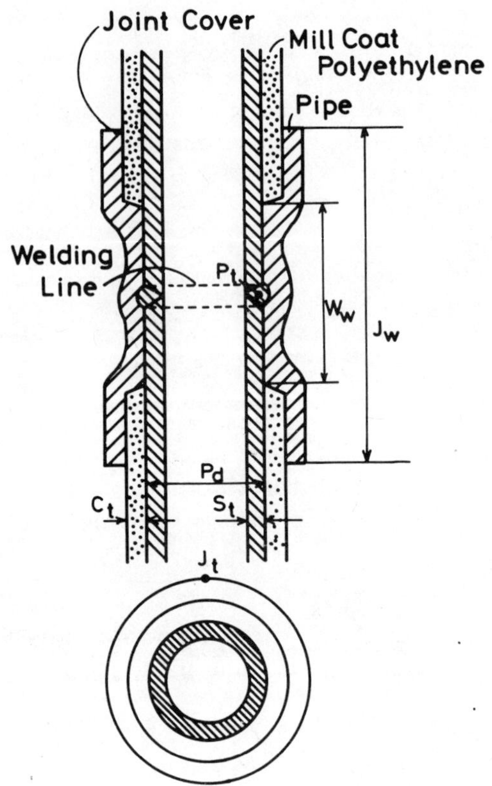

Fig. 1   Schematic section of welding part of pipeline and joint cover.

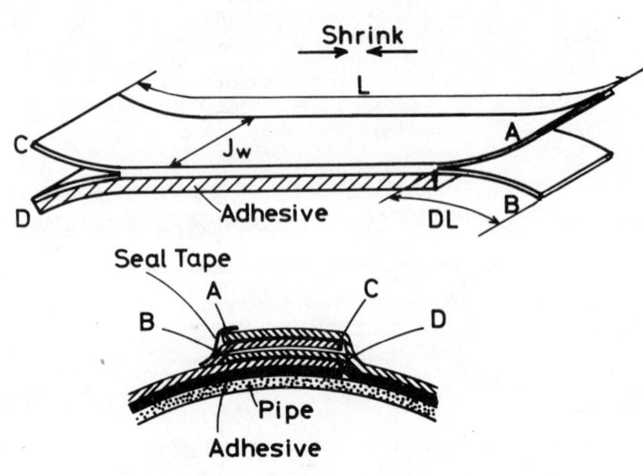

Fig. 2   Composition of a torch type joint cover (upper) and the engaged manner of both sheet ends (lower).

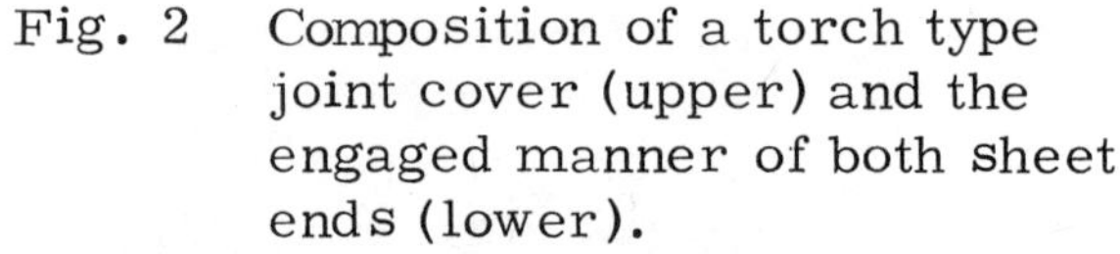

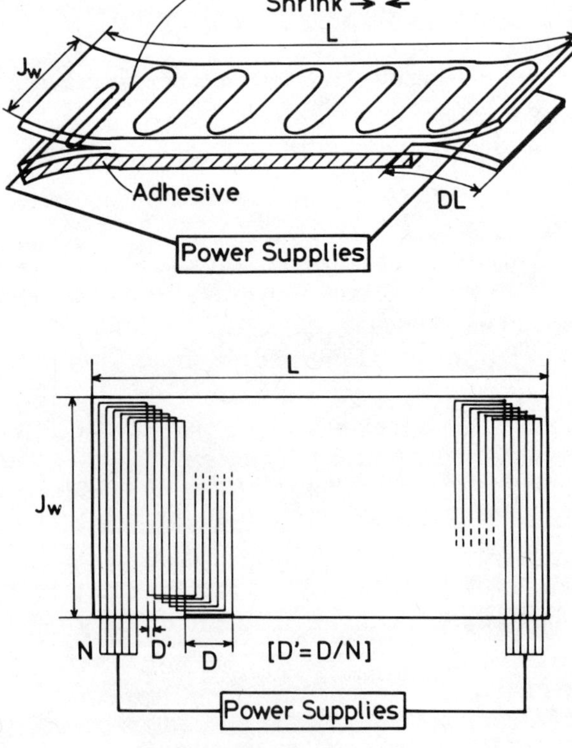

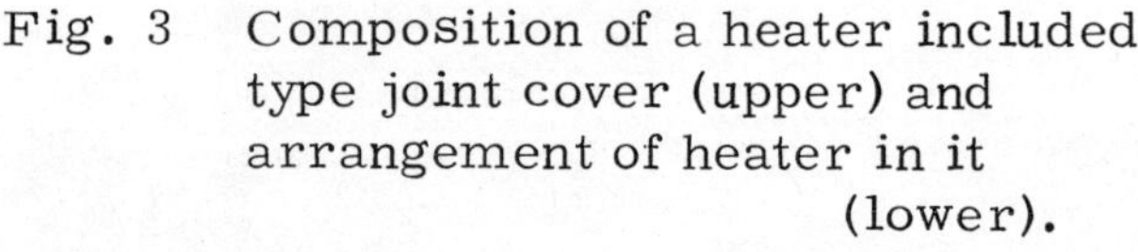

Fig. 3   Composition of a heater included type joint cover (upper) and arrangement of heater in it (lower).

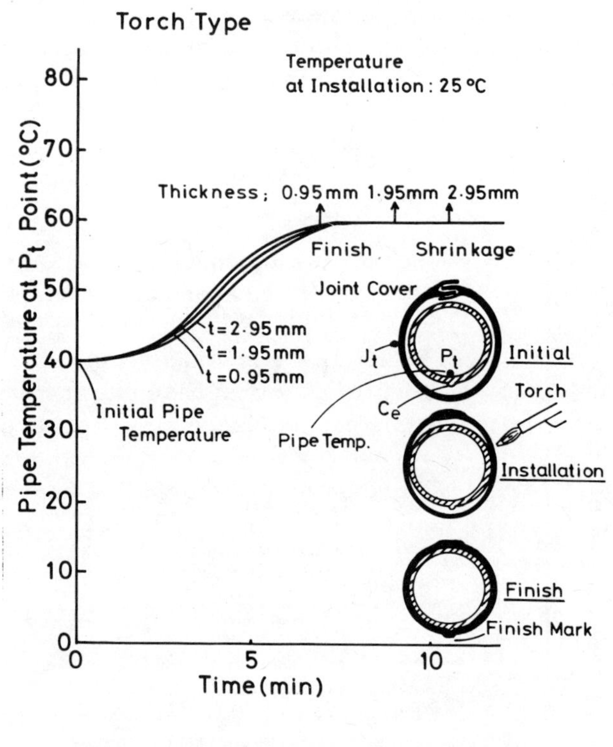

Fig. 4   Pipe temperature during the installation of torch type joint covers.

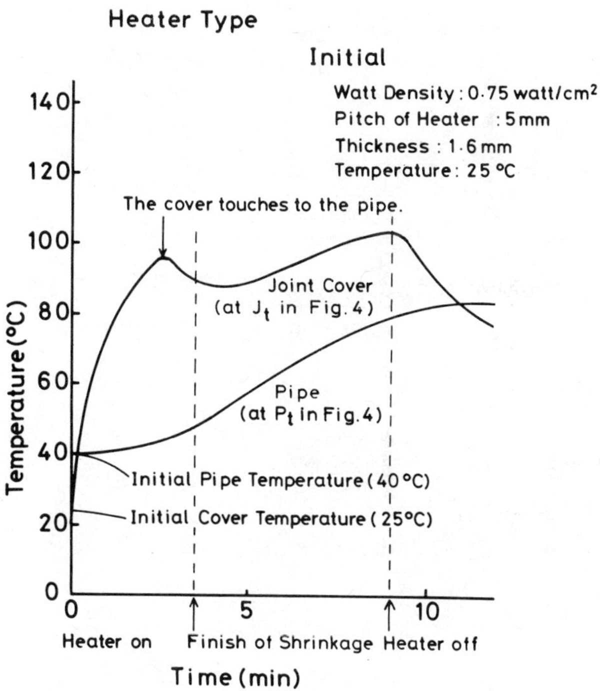

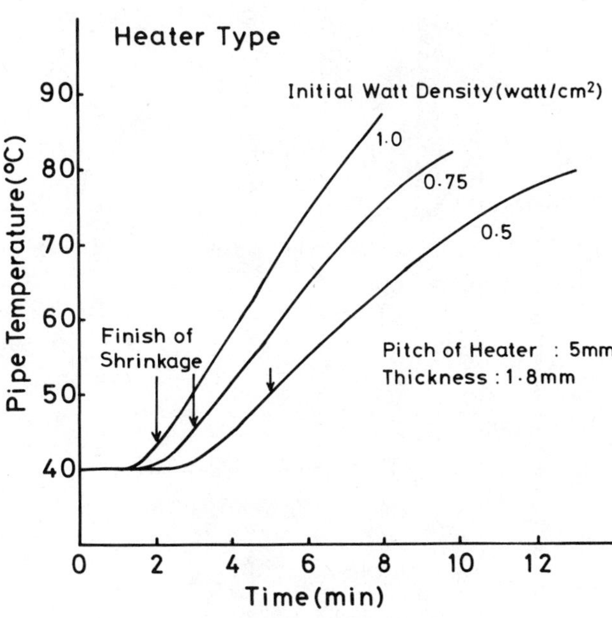

Fig. 5 Pipe and joint cover temperature during the installation of heater included joint covers.

Fig. 6 Pipe temperature of heater included joint covers with long time voltage application.

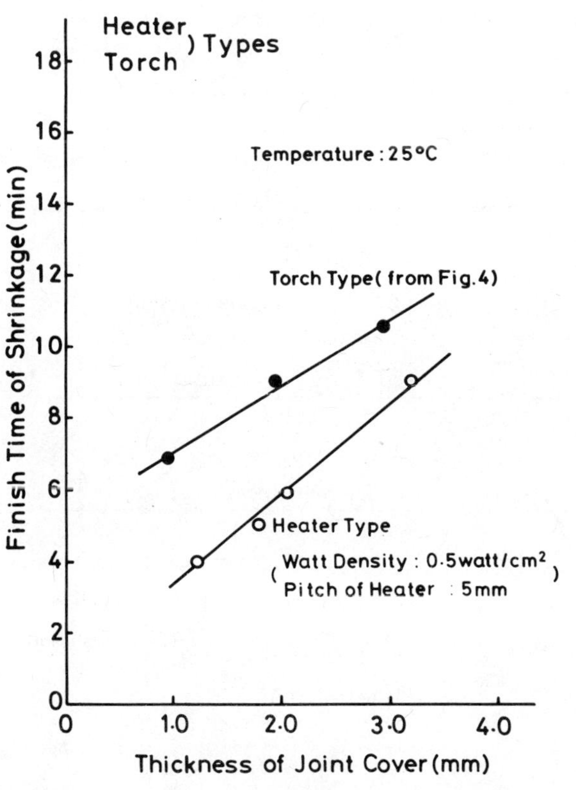

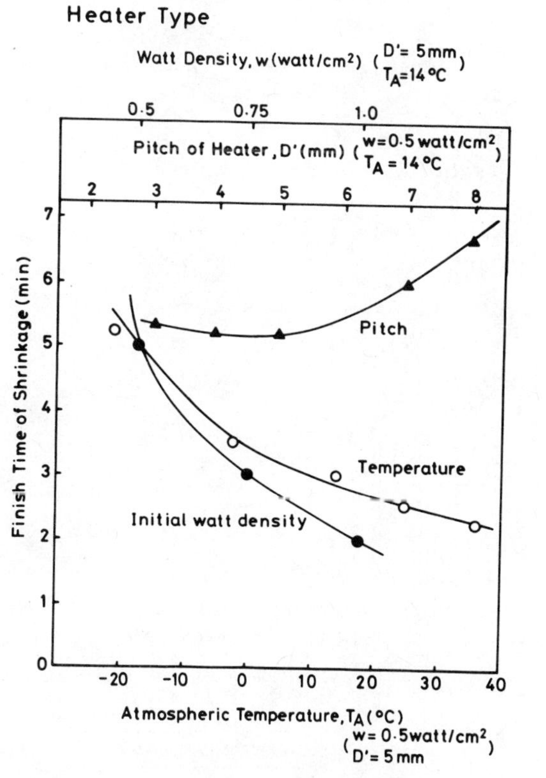

Fig. 7 Comparison of finish times of installation of torch type and heater included type joint cover.

Fig. 8 Finish time dependency of heater included type joint covers on various parameters.

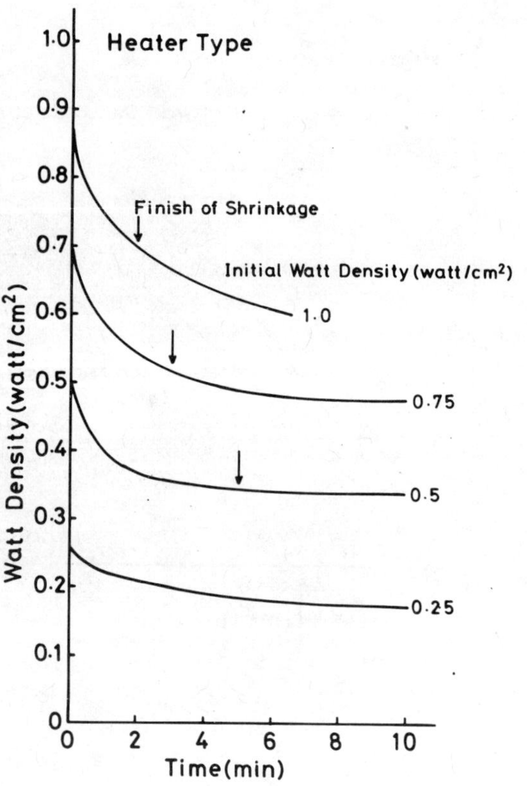

Fig. 9 Variations of watt density with installation time.

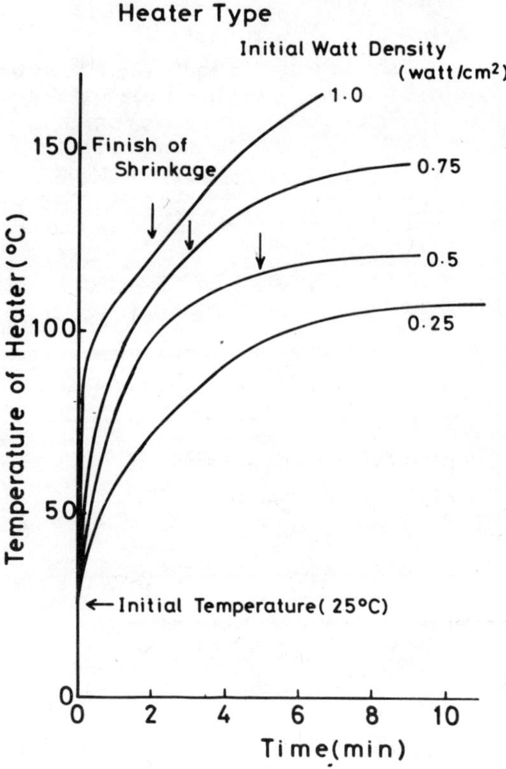

Fig.10 Temperature of heater surface.

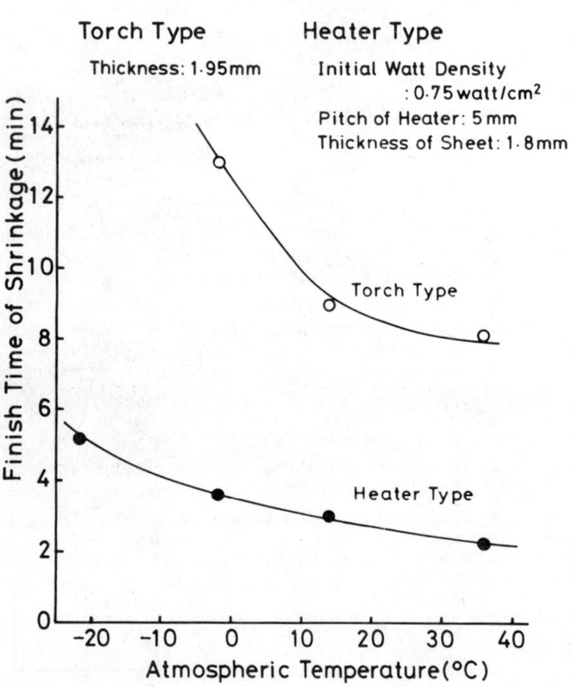

Fig. 11 Finish times of two tape joint covers under various air temperature.

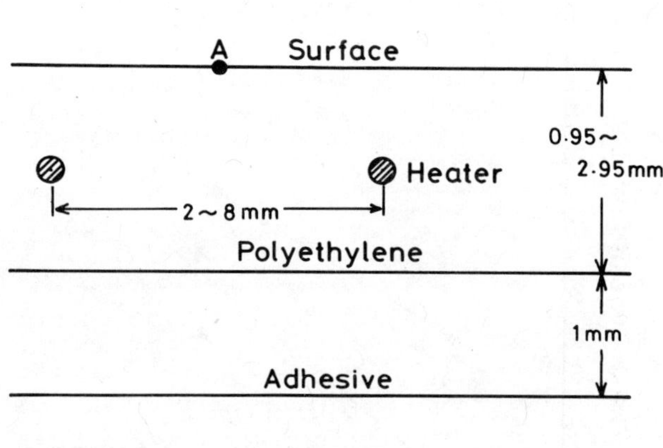

Fig. 12 Co-ordinates of a joint cover section for the computor simulation.

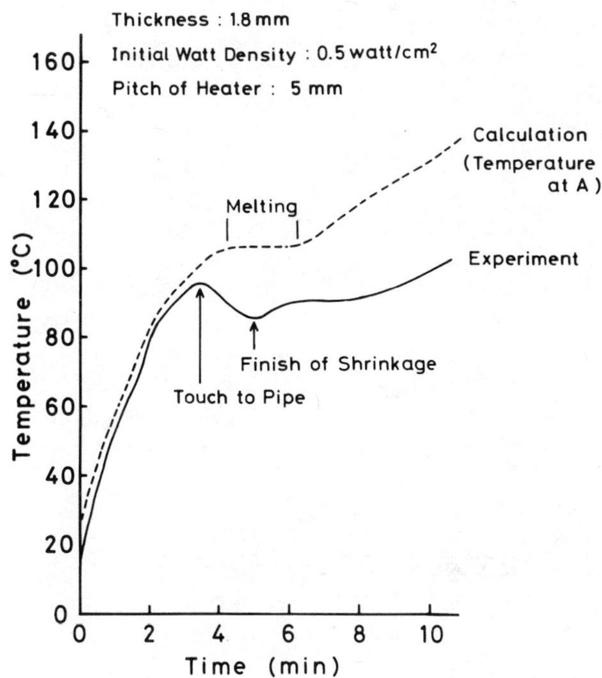

Fig. 13 Comparison of joint cover temperatures of experimental and calculated.

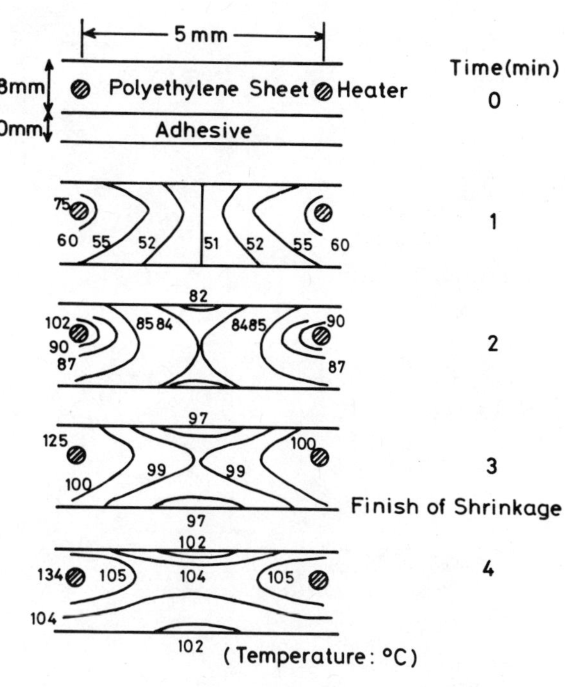

Fig. 14 Calculated temperature distribution in a joint cover with time.

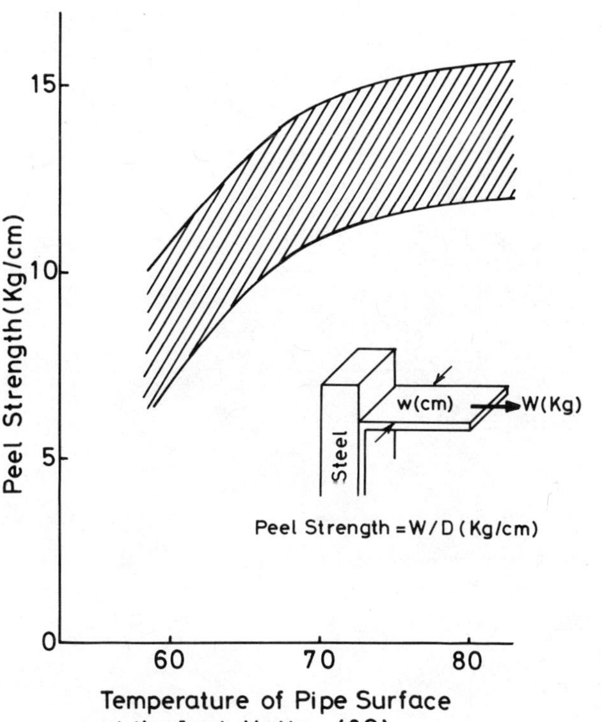

Fig. 15 Peel strength at 25°C of joint covers and steel installed at various temperatures.

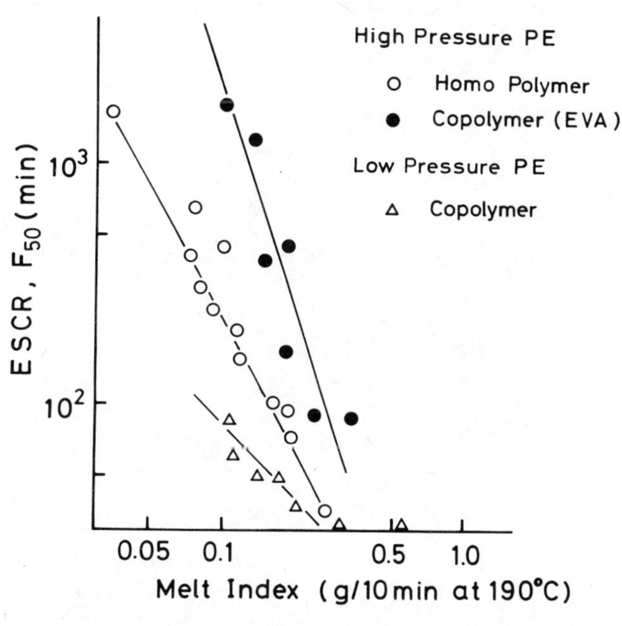

Fig. 16 ESCR of various types and melt flow index polyethylenes.

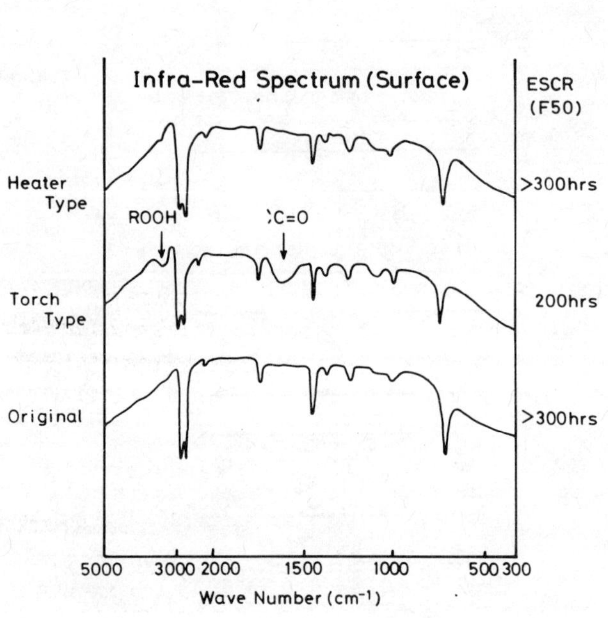

Fig. 17 Infra-red spectrum of installed two type joint covers and original raw polyethylene.

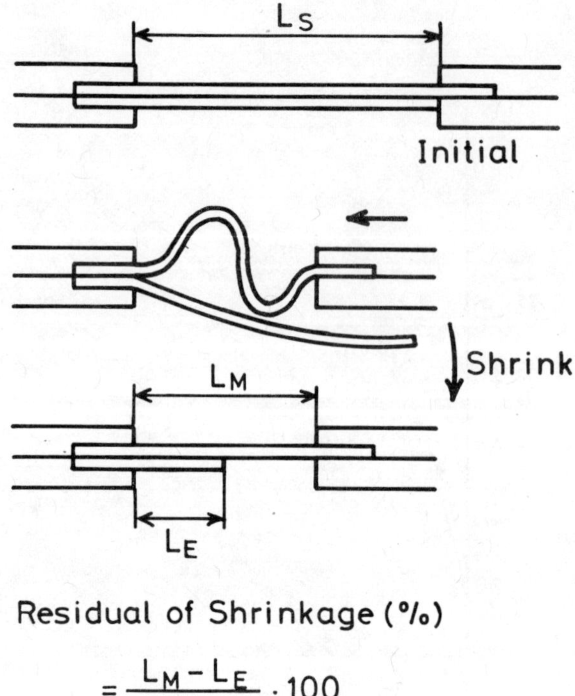

$$= \frac{L_M - L_E}{L_S} \cdot 100$$

Fig. 18 Definition and calculation of residual shrinkages of polyethylene sheet.

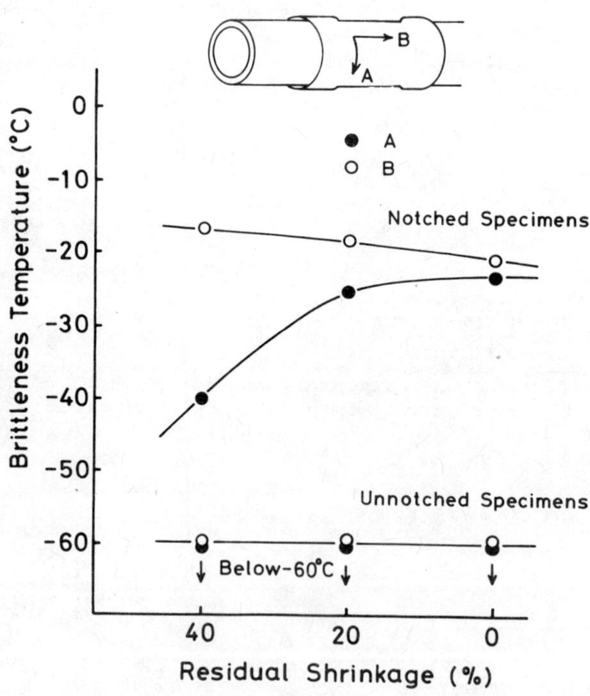

Fig. 19 Low brittleness temperature of various degree of residual shrinkages polyethylene.